The Geopoetics of Modernism

University Press of Florida

Florida A&M University, Tallahassee
Florida Atlantic University, Boca Raton
Florida Gulf Coast University, Ft. Myers
Florida International University, Miami
Florida State University, Tallahassee
New College of Florida, Sarasota
University of Central Florida, Orlando
University of Florida, Gainesville
University of North Florida, Jacksonville
University of South Florida, Tampa
University of West Florida, Pensacola

The Geopoetics *of* MODERNISM

Rebecca Walsh

University Press of Florida

Gainesville | Tallahassee | Tampa | Boca Raton

Pensacola | Orlando | Miami | Jacksonville | Ft. Myers | Sarasota

Printed in the United States of America on acid-free paper

This book may be available in an electronic edition.

20 19 18 17 16 15 6 5 4 3 2 1

Library of Congress Cataloging-in-Publication Data
Walsh, Rebecca Ann 1970– author.
The geopoetics of modernism / Rebecca Walsh.
pages cm
Includes bibliographical references and index.
ISBN 978-0-8130-6051-4
1. Modernism (Literature)—United States. 2. American poetry—20th century—History and criticism. 3. Geography in literature. 4. Geographers—United States. 5. Huntington, Ellsworth, 1876–1947. 6. Semple, Ellen Churchill, 1863–1932. 7. Stein, Gertrude, 1874–1946. 8. H. D. (Hilda Doolittle), 1886–1961. 9. Whitman, Walt, 1819–1892. 10. Hughes, Langston, 1902–1967. 11. National geographic magazine. I. Title.
PS228.M63W35 2015
810.9'112—dc23 2014031888

The University Press of Florida is the scholarly publishing agency for the State University System of Florida, comprising Florida A&M University, Florida Atlantic University, Florida Gulf Coast University, Florida International University, Florida State University, New College of Florida, University of Central Florida, University of Florida, University of North Florida, University of South Florida, and University of West Florida.

University Press of Florida
15 Northwest 15th Street
Gainesville, FL 32611-2079
http://www.upf.com

Contents

Figures

Acknowledgments

First, I would like to thank the anonymous readers at the University Press of Florida for their insightful and careful feedback. Acquisitions editor Shannon McCarthy warrants special mention for her enthusiasm for this project and for her incredible skill and attentiveness. I am also grateful that the manuscript benefited from Nevil Parker's and Elaine Durham Otto's editing dexterity.

For archival research assistance, I thank Elizabeth Dunn at Duke University's David M. Rubenstein Rare Book and Manuscript Library and Nancy Kuhl and Natalia Sciarini at Yale's Beinecke Rare Book and Manuscript Library. For permission to reprint archival material and visual images, I thank the estates of Langston Hughes, H.D., Bryher, and Charles Addams. For permission to reprint extensive quotations from Gertrude Stein's *The Geographical History of America,* I thank Random House. (Any third party use of this material, outside of this publication, is prohibited. Interested parties must apply directly to Random House LLC for permission.)

Susan Stanford Friedman's encouragement and guidance have been invaluable to this project. Her depth and breadth of knowledge, intellectual generosity, and commitment to mentoring young scholars continue to be unparalleled. Priscilla Wald, Susan Walsh, Laura Severin, Celena Kusch, Jody Cardinal, Kathleen Kearns, Anne Baker, and Jennifer Ho have been generous with their time and acuity in reading portions of the manuscript, and I thank them. Scott Kirsch's help with my navigation of historical geography is much appreciated. For research assistance and help with the final manuscript, I gratefully acknowledge Lindsay York and Kristina Bender.

At the University of Wisconsin–Madison, my work with Bruce Burgett, Mary Layoun, and Rob Nixon helped me refine my ideas. Cyrena Pondrom deserves special thanks for her eye-opening introduction to the world of modernist poetry as does Lynn Keller for exposing me to an exciting array of later experimental work. For intellectual exchange and friendship, I thank Andrea Kaston-Tange, Michelle Ephraim, Mara Scanlon, Susan Koenig, David Charbonneau, Celena Kusch, Sue Uselmann, Rebecca Schoenike Nowacek, Alicia Kent, and Tisha Turk.

Jody Cardinal deserves special recognition for her continuing friendship and essential long-distance exchanges of work and ideas.

At Southern Illinois University–Edwardsville, I thank Charles Berger, Catherine Selzer, Howard Rambsy, and Susan Fanetti. At Duke University, colleagues and friends provided valuable feedback as I shaped my book: Rob Mitchell, Sean Metzger, Lauren Coats, John Miles, Matt Cohen, Bart Keeton, Diana Solomon, Matt Brim, Jason Mahn, Kristin Solli, James Thrall, Denise Comer, and Priscilla Wald. I am fortunate that Joe Donahue, Michael Moses, and Joe Harris contributed to this lively intellectual community.

I am grateful to the English Department at North Carolina State University, particularly to Tony Harrison, for material support for my research, often at precisely the right moments. Here and in nearby universities, I am lucky to count among my colleagues and friends Jennifer Ho, Ora Gelley, Ariana Vigil, Laura Halperin, Jordynn Jack, James Mulholland, Anne Baker, Laura Severin, Leila May, John Charles Williamson, Sarah Sharma, Mark Dudley, Sharon Joffe, and the honorary Durhamite, Mark Sheftall.

I thank my parents, Robert and Jean Walsh, for their lifelong passion for reading and ideas. I only wish they could have seen this book in print. The intellectual generosity of Susan Walsh and Douglas Lanier has helped me as a researcher, writer, teacher, and colleague in countless ways. I also owe my sincerest thanks to Kevin and Tracy Walsh and to all the Shahs, especially Mary, for support and humor. Demonstrating the idea that family is not limited by biology or law, Claudia and Nestor Ramirez deserve special appreciation for their thoughtfulness, generosity, and care.

I am very glad to be able to share this book with James Shah. I am grateful to him—for his love, his unshakable faith in my work, and his unmatchable sense of humor—and to our son, Eamon Finn, who is a complete joy (and whose spatial instincts already rival mine).

Abbreviations

AHGC	*American History and Its Geographic Conditions* (Ellen Churchill Semple)
CP1	*The Collected Poems of Langston Hughes*
CP2	*Collected Poems, 1912–1944* (H.D. [Hilda Doolittle])
EA	*Everybody's Autobiography* (Gertrude Stein)
GHA	*Geographical History of America* (Gertrude Stein)

Introduction

Geographical Encounters, Modernist Geopoetics

Stephen Dedalus
Class of Elements
Clongowes Wood College
Sallins
County Kildare
Ireland
Europe
The World
The Universe (Joyce 12)

Featured in the early pages of James Joyce's modernist novel *Portrait of the Artist as a Young Man*, this is a list that Stephen Dedalus has composed on the flyleaf of the geography textbook he uses for his "Elements" class at Clongowes school. Stephen's relationship to this geography textbook becomes an occasion for him to work through recurring questions of Irish identity and belonging that are spurred by his family's debates about Irish national politics, British colonialism, and the place of Irish Catholicism in Ireland's political life as well as by his alienating experiences at his Jesuit boarding school. He uses it, significantly, to generate ever-widening concentric circles of belonging that serve to define and stabilize his identity in the face of instability. Joyce's novel explicitly evokes the disciplinary practices of academic geography, going so far as to typographically inset this passage apart from the rest of the narrative, in the form that a geography exercise would conventionally take. When read in the context of the dominant strains of academic discourse prevalent during Joyce's time of writing, Stephen's list takes on the performative aspect of environmental determinism at work. After he writes out the various spatial contexts in which he can locate himself, he does a curious thing in rereading the list backwards, so that environment *precedes* identity: "Then he read the flyleaf from the bottom to the top till he came to his own name. That was he" (13). In this moment, the universe, the world, Europe, Ireland, County Kildare, Sallins, his school—all of them external, environmental coordinates—conspire to produce Stephen Dedalus as a subject. The next to last item in his reverse reading is his "elements" class, suggesting that his academic classroom, and the geographical epistemology it teaches, most immediately and intimately produce his identity.

Although Joyce's considerations of geography have been discussed by scholars, this passage's relationship to environmental determinism has not.[1] And yet this idea that environment operates as a controlling feature of human society was, according to many historical geographers, *the* core idea in academic geography during this era, as the discourse of environmental determinism had taken hold as a widespread touchstone in academic geography in much of Western Europe and the United States during the late nineteenth and early twentieth century. In this particular brand of geographic epistemology, environmental factors ranging from topography to climate were thought to play a major role in influencing the organization and structure of human societies, a sort of environmental molding that left little room for human agency to determine social organization and values.

These details of Joyce's novel signal the not always easily noticed but nevertheless robust connections between literary modernism and historical forms of geographical thinking that *The Geopoetics of Modernism* takes as its central focus. As I will discuss shortly, they also subtly point to the role that poetry can play within this matrix. Despite the disconnect we might assume between literary avant-garde modernist practices and the ostensibly "dry" or seemingly remote work of academic geographers, the connections between these areas of cultural production were more commonplace, more varied, and more transnational than we might typically assume. My aim in this book is to reconstruct the conditions of reading, influence, and intervention in which literary modernism and geographical knowledge co-evolved, which can help us to see connections between a range of American modernist poets and the work of well-known American academic geographers such as Ellsworth Huntington and Ellen Churchill Semple as well as the popular "middlebrow" geographical periodical, the *National Geographic Magazine*.[2] This book traces these connections, reading the experimental poetic modernism of Walt Whitman, Langston Hughes, Gertrude Stein, and H.D. in relation to both academic and popular forms of geographical production. Focusing attention on poetry in particular provides a powerful case study that attests to the breadth of cultural materials that participated in broader negotiations during the modernist period about the nature of the relationship between the physical environment and the human.

The contours of particularly American poetic encounters with geographical epistemologies are deep and wide. They take their origin in Walt Whitman's engagements with the European geography of Alexander von Humboldt and Mary Somerville, and they reflect in the modernist period the singularly strong tradition of environmental determinism in American academic geography as well as the unique relay between it and the mainstream articulations of the *National Geographic*.

While a combination of academic and middlebrow popular sources introduced particular geographic epistemologies into American culture, modernist poets responded to these ideas but did not necessarily produce American geography as content or reproduce the nation in geopolitical terms. Instead, geographic sources enabled, and made legible, a centrifugal poetic production of global geography and the adoption of transnational perspectives. Academic and popular geography provided the grounds on which modernism could use geographical location as a synecdoche for cultural and national identity; its interest in global connection and comparison presented modernism with formal logics of spatial comparison, homology, and transitivity capable of bringing together disparate geographical locations from distant global addresses. Geography and modernism can thus be seen not only as structural analogues for each other but as multiple voices contributing to a common conversation about the nature of the relationship between society and the environment, about how we might navigate units of scale (region, nation, transnation, globe), and about the cultural, national, and racial politics of how we orient ourselves on the globe.

The modernist poets I consider here—Walt Whitman, Langston Hughes, Gertrude Stein, and H.D.—can be linked by biographical/archival materials as well as by cultural context to these globally focused traditions of geography. While they have not routinely been at the forefront of either the "spatial turn" or the "transnational turn" embraced by the New Modernist Studies of recent years, they all ground their texts, to varying degrees, in the particulars of global place-names, landscapes, and topographies that defy the limits of a single national territory.

Unlike the wealth of traditional scholarship that has examined modernism's experiments with time, temporality, history, and myth, my focus is less oriented by the *anachronistic* elements of modernist poetic experimentation, in which time is out of place, than by what we can call the *anachoristic* attitude of these texts, in which place is "out of place."[3] The writers I consider all deploy to varying degrees formal strategies of parataxis that "mix up" conventionally distant and culturally, nationally, or racially different locales on the globe. Near and far, same and different, and self and other collapse and become realigned in Whitman's vision of a globally connected landscape in "Passage to India," Hughes's constructions of an African diasporic geography in his lyrics of the 1920s, Stein's comparative, airplane's-eye view of America in *The Geographical History of America*, and H.D.'s hybridized landscape of London and Egypt in *Trilogy*. In embracing global locations in ways that eliminated the distance between the near and the far, modernist poets thus problematized the boundaries of nation, of "home" versus "foreign," colonial epicenter versus colony, "West" versus "non-West," geopolitical ally versus enemy.

Given the global geographic content these writers take on, parataxis within their poetry operates much like the global comparison of geography only scaled at the level of form and pushed in a variety of directions. The transnational politics these kinds of moves generate remain open on at least some level. According to canonical discussions of parataxis, this sort of juxtapositional formal strategy resists conventionally stable and fixed linguistic connections, generating, as Theodor Adorno has articulated, "artificial disturbances that evade the logical hierarchy of a subordinating syntax" (131). For him parataxis is "critical and utopian," an "anti-principle" that works against dominant systems of meaning (115, 140). It is unstable, ambivalent, and contradictory, shifting across established hierarchies and circuits of power. In the landscape of modernist poetry I map here, we see American poets generating open, comparative forms that write against not only the fixative, binaristic discourses of nationalism, imperialism, colonialism, and racism but also some of the implicit ways that environmental determinism itself encoded American exceptionalism and the imperial superiority of the West. Modernist poetry becomes a form of geography, in a sense, in which traditional geopolitical alignments are complicated and redrawn.

The concept of geopoetics signaled by the book's title, then, refers in part to the geopolitical work performed by modernist poetry when it "mixes up" conventional geography and the racial, cultural, and national divisions and hierarchies that attach to it. These gestures have the potential to undermine the foundations of American exceptionalism, primitivism, and orientalism, although they are not entirely free of the darker side of such difference discourses, either. If the "geo" in the term *geopoetics* stresses the global geographical content of the modernist poetry represented here as well as its engagements with the geographical epistemologies in the culture at large, the second part, "poetics," signals the potential for poetic experimentation to create incommensurate and sometimes self-contradictory political positions regarding ideas of cultural, racial, and national otherness. Part of this book's argument asserts this geopolitically uneven terrain of American global poetry, produced by open poetic form. The other part argues that the geopoetics of modernism generated a significant counterpunch to the ontological fixity of geography itself that environmental determinism took as its starting point.

Recognizing the geopoetic anachorism exemplified by Whitman, Hughes, Stein, and H.D. helps to create a more visible place in the "New Modernist Studies"[4] for the transnational and global dimensions of poetry in general and for experimental poetics in particular. Well into the "transnational turn" that has expanded the spatial boundaries of knowledge projects in the humanities and social sciences, the almost exclusive focus on narrative among scholars in American Studies and Modernist Studies has not shifted very far in considering other literary genres.[5] A brief glimpse at recent work that conceptualizes cosmopolitanism, imperialism, and globality or its concerted disavowals reveals a terrain heavily dominated by the novel, with occasional forays into memoir and short fiction.[6] Jahan Ramazani's work stands out as a notable exception to the narrative-heavy focus of much recent scholarship, exposing some of the long-standing assumptions about poetry's purportedly narrow, umbilical rootedness, articulated by modernist poets such as Eliot and Auden as well as by recent scholars

("Poetry, Modernity, Globalization" 290)[7] and advancing instead a strong case for the opposite axiom: that modernist poetry, perhaps above all other literary forms, contains "nation-crossing" within its very DNA (298). While Ramazani's work relies heavily upon the anthropological work of James Clifford and postcolonial theories of hybridity, my design here is to shed light on the historical epistemological conditions of the modernist period—the underlying intellectual formations and the forms of disciplinary authority that attach to them—that enabled and/or constrained the border crossings visible in modernist poetry.[8]

Poetry is a genre that has historically engaged with a full range of cultural discourses, materials, and practices, whether high or low, even though it has all too often seemed to lag behind narrative and other objects of analysis that more frequently attract the interest of cultural studies scholars.[9] Two factors may explain why poetry has not been fully integrated into cultural studies work. Adelaide Morris, among others, has noted that the "aesthetic autonomy" ascribed to modernist poetics, a holdover from romanticism, "takes poetry apart from the market exchanges through which it circulates and the historical debates in which it participates" (2).[10] And yet when poetry is freed from the *cordon sanitaire* of "aesthetics," there can be an equally problematic tendency to search for sociohistorical elements as though poetry were merely a reflection, trace, or inert container, a practice that gathers poetic evidence in order to confirm the sociohistorical.[11]

Lessons about poetry's ability to intervene in rather than merely reflect cultural conversations dominant in the modernist period appear, oddly enough, within the same scene discussed above in *Portrait of the Artist as a Young Man,* one of modernism's most canonical novels. That scene stages a face-off of sorts between academic geography and poetry, the same set of negotiations that *The Geopoetics of Modernism* takes as its main focus, and it highlights the ways in which these two seemingly disparate sites of knowledge each weighed in, in overlapping and often conflicting ways, about how social bodies might define themselves and various social others in relation to geography. Next to Stephen's geographic list, his classmate Fleming has given Stephen a ribbing by writing a parody of it, "for a cod":

> *Stephen Dedalus is my name,*
> *Ireland is my nation.*
> *Clongowes is my dwelling place*
> *And heaven my expectation.* (13)

Fleming's children's rhyme demonstrates the power of poetic, non-narrative forms to transform the geographical deterministic epistemologies bound up in dominant conventions of academic geography and make them useable for other ends.[12] The first line, "Stephen Dedalus is my name," and its line break allow Stephen to possess his own identity independent of whatever spatial coordinates follow. The spatial frames that do unfold in the poem violate the principles of linear spatial scale, as the national (Ireland) appears before the local (Clongowes), which appears before the largest and most diffuse kind of spatial scale, heaven. Finally, Ireland, Clongowes, and heaven are all for Stephen "my" objects of orientation, identification, and desire that are external to him in ways that he can then overtly possess or reject. At the same time that Fleming's poem implicitly critiques Jesuitical regulation of its subjects, he lends a sense of agency and ownership missing in Stephen's own geographic list. Poetry loosens the lockstep tracery of linear spatial scale and environmental determinism as encoded by academic geography. Significantly, Stephen recognizes that he himself has generated a list but that Fleming has generated poetry: "He read [Fleming's] verses backwards but then they were not poetry" (16). Though this is a children's rhyme written to be playful, it is nonetheless poetry that prompts Stephen to reevaluate his own geography exercise and to contemplate "big thought[s]" like the nature of "everything and everywhere" (13).

Tracing modernism's overlooked poetic engagements with and interventions in dominant forms of geographical thinking requires combining cultural materialist and historicist methodologies with considerations of both poetic form and transnational politics. In this sense, the analysis that follows participates in modernism's "transnational turn" as well as its recent turn toward the spatial, with the intent to secure in both scholarly conversations a more prominent place for poetics and for historicist attention to the dominant geographic epistemologies that circulated during modernism's heyday. Recent spatial reorientations in modernist studies have responded to the clarion call of cultural geographers such as Edward Soja who warn that because of the "apparently innocent spatiality of social life," space, more so than time, may hide the consequences of power and ideology (*Postmodern Geographies* 6). This has meant a paradigm shift away from decades of modernist scholarship devoted to temporally inflected concerns such as Bergsonian notions of inner time, psychoanalytic developmental processes of identity formation, disruptions in or frustrations of narrative teleology, or, at the metadiscursive level, periodizing exactly when modernism began and when, and if, it ended. This shift, I might add, liberates us from the sort of "philosophical miserabilism," as Doreen Massey calls it, which has sometimes accompanied discussions of temporality because of its associations with the fear of death (58).

Recent projects have instead focused on the ways that spatiality shapes literary responses to modernity on a number of levels. In redrawing the map of modernism's geographical nodes of production, we are moving beyond the familiar New York–London–Paris emplotments and toward a more varied and broad-ranging worldscape that includes the suburban, the rural, the non-European, and the socially marginalized, a multiplication of the geographical "margins" of modernism-as-usual, as the essays in Laura Doyle and Laura Winkiel's *Geomodernisms* demonstrate. This reimagined terrain of modernism's production—"the geographies of modernism"—means we are no longer asking "when was modernism" but "where was modernism."[13] Henri Lefebvre's attention to the dialectical nature of the production of space and the production of social relations under global capitalism (*Production of Space* 85) has informed numerous spatialized readings of the ways that modernist literary texts reflect, reproduce, and challenge social space through their representations of the topographies, landscapes, and architectures of modernity. These kinds of readings help to constitute in significant ways what Peter Brooker and Andrew Thacker have called modernism's "critical literary geography" (5). Such developments owe a partial debt to the emphasis placed by postcolonial and diaspora studies on issues of territoriality, displacement and emplacement, and the nature of cultural identity, and to recent attempts by "locational" feminist theorists and feminist geographers to circumvent the essentialist-constructivist impasse in identity politics by using space and location as an analytic.[14] In this regard, the essays in Brooker and Thacker's recent collection *Geographies of Modernism* are instructive in their efforts to interweave spatial readings of the content of modernist texts with sensitivity to their historical and geographical sites of production and with recognition of "the historical trajectory of space and geography informing modernism and modernity" (2).

There is a sharper distinction to be made, however, between the historical changes in space and geography in modernity at the center of Brooker and Thacker's project and the important role played by discourses *about* space and geography, epistemological practices that originated to a powerful degree in academic and mainstream geographic texts. For whatever reason, modernist scholarship has come fairly late to this latter form of understanding, especially when it comes to American modernism and to poetry in particular. Of the small number of recent contributions in modernist scholarship to consider the role of academic geography, the focus has been on Anglo-European modernist literature and Anglo-European geographers in particular. Perhaps not surprisingly, given the entrenched scholarly assumptions about poetry outlined above, such work has also exclusively focused on narrative.[15] American Studies scholars have been attentive for far longer to the production of geographic epistemologies in earlier periods of American literature, although most of this work displays a similar attention to narrative.[16] There is more therefore to be gained by examining early-twentieth-century American geography as well as the ways that American poetic modernism, and by extension American modernism generally, contended with dominant academic and popular discourses that shaped how geography could be perceived in the period. This is a fruitful way of attempting, as Edward Soja has sought to do, to "open up and recompose the territory of the historical imagination through a critical spatialization" (*Postmodern Geographies* 12). Among other things, focusing on these kinds of disciplinary and epistemological negotiations positions modernist literature as an active contributor in ongoing conversations about modernity rather than risking treating modernist literature as a passive reflector of modernity's spatial conditions and its modes of capitalist reproduction. At the very least it brings together Fredric Jameson's clarion call for humanities projects to "always historicize" with the more recent imperative to "always spatialize" that has followed in the wake of David Harvey, Doreen Massey, and Edward Soja's recent contributions to cultural geography.

Historicizing these conversations, particularly in an American context, necessarily involves contending with environmental determinism, despite the ways in which this paradigm has long been viewed in contemporary academic geography circles as a set of quaint, overly simplistic, and very outdated notions. It is significant that modernist writers were representing and in many cases radically remodeling global geographies at the very same time that environmental determinism emerged in America as a widespread and powerful paradigm. This model of geography was not limited to American universities, as it was articulated to varying degrees by a group of geographers that included German geographer Friedrich Ratzel, British geographer Halford Mackinder, and American geographers William Morris Davis, Ellsworth Huntington, and Isaiah Bowman. This is a large, diverse group, yet for English-speaking audiences on both sides of the Atlantic, environmental determinism was most readily identified with the American geographer Ellen Churchill Semple. This strain of geographic thinking in the late nineteenth and early twentieth century was heavily invested in several key ideas: that the features of the landscape and climate are the first stop in any sort of inquiry into human society, that the geographic environment influences human society (not necessarily the other way around), and that similar physical environments can produce similar cultural and societal outcomes, no matter the physical distance separating them. Semple's well-known *The Influences of Geographic Environment* (1911) begins with a statement that, although not fully reflective of her own uneven relationship to environmental determinism over the arc of her career, nevertheless makes a bold and highly influential set of claims that came to define this type of thinking:

> Man is a product of the earth's surface. This means not merely that he is a child of the earth, dust of her dust, but that the earth has mothered him, fed him, set him tasks, directed his thoughts, confronted him with difficulties that have strengthened his body and sharpened his wits, given him his problems of navigation or irrigation, and at the same time whispered hints for their solution. (1)

Like that of her peers, Semple's approach to the "man-land" equation was partly grounded in Lamarck's ideas of organic memory which stressed hereditary forms of adaptation to external pressures in the natural world. Geographers produced environmental determinism as their own scientific version of adaptation, partly as a response to geography's sense of disciplinary crisis in the late nineteenth century. In an effort to science itself up, so to speak, at a critical junction in the late nineteenth century of increasing disciplinary specialization, the field sought to establish itself as distinct from history or classics, two fields whose methods had long been blurred with geographical inquiry. It also attempted to establish its rigor by not only generating "hard," quantifiable data on topography, climate, landscape features, and geomorphology but also using that data to generate theories of cause and effect that had been identified with the scientific method, in this case positing specific environmental pressures as the cause and human social features as the effect.

Environmental determinist ideas also provided the West with a legitimation theory that could justify its dominance over other regions of the globe. The ethnic character of peoples inhabiting locations monotonously less "temperate" was seen as a direct product of the extremes of their environmental conditions. Their inability to be both properly "tempered" and properly stimulated by a seasonally or topographically variable environment was viewed, in turn, as an inability to be responsible stewards of themselves and whatever natural resources existed in their regions. Such ideas reconstituted and made systematic the links between environmental and social superiority expressed in the continental philosophy of Kant, whose *Geography* views environmental differences as the root cause of western superiority: hot lands and frigid lands beyond the West's temperate zone each produce lazy, timid, and superstitious people, from those in the tropics to the Ostoyaks, Samoyeds, Lapps, and Greenlanders in the extreme north.[17] Environmental determinism colluded with Social Darwinist ideas to produce the notion that the West could prove its global superiority by demonstrating its own adaptability as it applied itself and its own cultural formulas to seemingly extreme or remote corners of the world.[18] Approaches to geography in these early moments of western modernity were indeed far from the "innocent spatiality" Soja discusses, and the modernist poets in this project were not always immune from the ideologies of western superiority and American exceptionalism that such practices encoded.

In light of these tendencies, perhaps environmental determinist ideas are likely to surface whenever overt global inequities need to be understood or even explained away. Anyone assuming that this paradigm has been fully debunked and is outmoded need only notice the role that its central assumptions play in Jared Diamond's wildly successful 1997 *Guns, Germs, and Steel,* which won the Pulitzer Prize in history in 1998 and received widespread public attention in academic and popular media outlets. Diamond, initially trained as a physiologist, employs an interdisciplinary biogeographical method in comparing the relative environmental advantages that led to various Western European successes on the global stage over the last 13,000 years. His argument begins with the premise that "human development on some continents got a head start in time over developments on others," with his book ultimately offering a master narrative about the "effects of continental environments on history" (28). Eurasia, in Diamond's model, benefited from a more rapid East-to-West spread of wheat cultivation and domestication of animals because of the similarities of climates roughly along the same latitude and because of the temperateness of all of its regions. By contrast, Africa and the Americas' greater climatic variability, given their North-South orientation, hindered this kind of spread, as did their relative lack of temperate environments. According to this view, Eurasia's longer and more robust cultivation of domestic animals led to the development of diseases—and resulting forms of population resistance—from millennia of close interaction with various animal disease-producing hosts (the germs of Diamond's title). Agriculture and livestock were also the key to creating densely populated regions that needed hierarchical governments that were centrally organized, capable of generating a surplus of food able to sustain people beyond the food producers themselves: artists, artisans, blacksmiths, a military, etc. (guns and steel).

For Diamond, it is precisely these relative advantages that allowed the West to achieve global dominance relative to regions now in the so-called Third World, advantages that he traces directly to the controlling effects of the physical environment. His desire to chart these historical patterns of environmental "giftedness" at the broadest levels roughly follows the axiom, in one reviewer's words, "them that has, gets" (J. Terrell 9). These moves become less surprising, however, when we recognize that Diamond's polymathic scope and interest in the sort of expansive temporal sweep that Ferdinand Braudel would call the *longue durée* resuscitates the reach and orientation of late nineteenth- and early twentieth-century geographic determinism. Given the recent widespread popularity of Diamond's ideas, we can more easily imagine the ways that, in its early twentieth-century heyday, sometimes obvious and sometimes subtle versions of environmental determinist discourse found their way into popular periodicals like the *National Geographic* and its imitator, the British gazette *Countries of the World*, which aimed to both educate and entertain broad readerships, or the ways that modernist poets might have been aware of this discourse and were capable of forming reactions to it.

Recognizing the coevolution of literature and environmental determinism reframes modernist negotiations of difference discourse in new ways. Part of this involves recognizing how representations of landscape were put to overdetermined ideological and political uses within a climate in which such representations were saturated with ideas of cultural superiority and inferiority and imbued with a policing function that distinguished the West from "the rest." In one sense, modernist writers representing forms of racial, cultural, and national difference through the landscape were liberated from hardened, biologized ideas attached to the body that had been ushered in by Herbert Spencer and other "scientists" of race in the nineteenth century. Ellen Churchill Semple self-consciously departed from these notions, visible in her mentor Friedrich Ratzel's earlier volume *Anthropogeographie,* with an ostensibly democratic shift away from ideas of race and toward the effects of the environment, as I discuss in chapter 1. But given the weight of the ways that landscape was often leveraged in the modernist period against ideas of civilizational superiority and inferiority, modernist poets were nonetheless navigating operations of racial, cultural, and national hierarchies through their constructions of landscape, even when they were not overtly portraying racially, culturally, or nationally marked speakers or figures in their works.

Although the links between environmental determinism and the operations of imperialism are indeed a key way that environmental determinism has operated in modern geographic and scientific history, there is a danger in reducing all of its central assumptions and methods to some of its own particular conclusions and to their resulting social and geopolitical uses. Close reading of the environmentalist work of geographers like Semple makes visible some of the paradigm's key methods of global comparison: drawing analogies and homologies linking geographically distant but climatically similar locales, and advancing ideas of landscape transitivity that such comparisons instantiate. These are global, comparative tactics that drew together different and disparate geographies and cultures irrespective of national or geopolitical boundaries, strategies which take similar shape in the *National Geographic,* a magazine that was wildly popular in the early decades of the twentieth century and was read by a wide range of modernist writers.

Elizabeth Bishop's poem "In the Waiting Room" provides a well-known illustration of the sorts of global comparisons that the *National Geographic Magazine* invited and that modernist poetry pursued. Bishop's 1971 poem portrays a young girl who encounters an issue of the *National Geographic Magazine* while she waits for her aunt in a dentist's office. The photos inside the magazine depict "[b]abies with pointed heads" and "black, naked women" with "necks / wound round and round with wire" and "horrifying" breasts, images that prompt the speaker's explorations of cross-national, cross-cultural, and cross-racial identification. This becomes for her a part of her vertiginous loss of individual identity: "I—we—were falling, falling, / our eyes glued to the cover of the *National Geographic,* / February, 1918" (Bishop 4–6). The speaker considers the possibility that some of the similarities that possibly "held us all together / or made us all just one" are shared with the naked African women in the magazine (7). The poem's concluding focus on the speaker's immediate environment, a cold, wet winter evening in Worcester, Massachusetts, takes on an implied recontextualized meaning that the magazine provokes. Although Bishop's poem responds in part to the magazine's reputed use as a socially sanctioned vehicle for sexual fantasy given its regular features of "exotic" naked bodies, something expressed in more explicit terms in William Carlos Williams's *Paterson I*, it fundamentally illustrates not only the ubiquitous and familiar aspects of the *National Geographic Magazine*, childhood memories of which Bishop used to construct her poem, but also the implicit ways the magazine invited global connection and comparison.[19] Poetic engagements with the magazine visibly flower in Bishop's 1971 poem, but they existed in the modernism of the early decades of the twentieth century as well.

My argument stops short of claiming that modernist writers were themselves environmental determinists. Rather, we can locate in environmental determinist texts a structural analog for what modernist writers thought they could do with geography, much as Frazer's 1890 comparative study of global mythology, *The Golden Bough: A Study in Magic and Religion*, proved instructive in modernism's development of a mythic method, as traditional narratives of modernism would have it. While environmental determinism may have helped to make legible certain formal structures of space-annihilating global conjuncturalism, modernist writers were not limited by its underlying assumptions. A central disjunction between environmentalist geographers and modernist writers lies in perceptions of the ontological status of geographic space. Academic geographers were exceedingly invested in its ontological stability and assumed that cultural or social effects could always be traced back to some sort of environmental control or set of pressures. This was the master narrative of early twentieth-century geography. By contrast, geographic space in modernist experimental texts appears porous, unstable, and sometimes even physically portable because of the ways that conventionally distant and different locations on the space of the page merge or combine into new configurations. Following from Foucault, Don Mitchell in his discussion of urban planning and public housing has observed the tendency within western thought to believe that "rational (and rationalized) landscapes are necessary for social order" (115). The new and messy forms of global spatial relation produced by Hughes, Stein, and H.D. undermine these assumptions through their "anachoristic" representational strategies. It could well be that the entrenched notions of ontologically rooted place that environmental determinist ideology encourages provide the conditions for modernist experiments with geography on the page, given that the "clearer the established meaning and practices of a particular place the easier it is to transgress the expectations that come with place" (Cresswell 103). Such poetically engineered anachoristic arrangements implicitly posed a challenge to the environment–human master narrative of environmental determinism and, secondarily to some degree, to the legitimation of imperialism and colonialism that environmental determinism underwrote.[20] What we find, then, is that modernist experimental poetic texts were in some ways indebted to environmental determinism but at the same time they undermined—whether self-consciously or not—some of its most fundamental assumptions.

Enlarging the orbit of modernism's interdisciplinary negotiations also broadens the cultural and temporal boundaries placed around environmental determinism within the field of cultural geography. Histories of the internal developments within academic geography, largely concerned with geography's institutional and governmental forms, typically mark the 1920s as the moment when environmental determinism was dislodged by other paradigms.[21] However, as I discuss in chapter 1, its ability to inform or trickle down to other media contributed to a continuing relevance. And while it is often suggested that the early twentieth century saw a widening gulf between academic geography and popular periodicals, articles in the *National Geographic* throughout the 1920s and 1930s continue some of environmental determinism's fundamental assumptions when they approach various global cultures first and foremost through their environmental circumstances and hence their environmental influences. In high school textbooks and in the media realm exemplified by popular periodicals, then, environmental determinism had quite a firm grasp on conceptions of the "man-land" relationship until well into the 1950s. These mainstream geographical outlets helped to codify what I call an "environmental style" that operates on the level of narrative patterning: landscape first, culture second. This particular style also operates through transforming academic ideas of environmental influence into a portable, tactical rhetorical strategy that the *National Geographic Magazine*'s writers attach to a variety of subjects. A number of its articles conscript this environmentalist style for American exceptionalist ends; their celebrations of American cultural and technological achievements as superior to the rest of the globe are advanced through environmentalist rhetoric that suggests that American culture has triumphed over the environment when other parts of the globe are still subject to its influence. In this regard, the

National Geographic Magazine returns belatedly to environmental determinism's earliest justifications of cultural domination in the form of its implicit and sometimes explicit support of imperialism and western superiority. The net result is the continued visibility of environmental determinism's fundamental logic of environmental influence over the social, even when it is put under erasure, and an increasingly flexible way of adapting those ideas aesthetically in ways that both enable and constrain the varied poetic projects that the following chapters discuss. This kind of attention to the middlebrow periodical and to modernist poetic engagements with environmental determinism contributes to recent work in cultural geography that has encouraged attention to the myriad public platforms through which geographic knowledge was disseminated, contested, and shaped, in ways similar to what it means to do the work of public history.[22] This interplay between geography and modernism necessarily presents a different and more detailed picture of exactly when and under what circumstances environmental determinism held sway.

To provide this project with its historicist footing, chapter 1 charts the shift in dominant geographical epistemologies from the cosmic natural geography in the mid to late nineteenth century of influential geographers like Alexander von Humboldt and Mary Somerville to the rise of environmental determinism in late nineteenth and early twentieth century in the work of Friedrich Ratzel, Halford Mackinder, Isaiah Bowman, and, most notably, Ellen Churchill Semple. While American Studies scholarship has focused on the significance of homegrown forms of academic and popular geography in the nineteenth century, this chapter makes a case for the enduring influence of Humboldt and Somerville, transatlantic imports who introduced Americans to ideas of global unity and to an international, rather than national, outlook. My argument locates a similar global sensibility in the work of Semple, who actively worked against the disciplinary boundaries separating geography from history and literature even while advocating for an early version of "the spatial turn" that would place environmental determinist principles at its center. The last portion of the chapter considers the *National Geographic Magazine,* which disseminated environmental content and an environmentalist style to a wider public that included modernist writers, translating those ideas into an adaptable aesthetic.

Though they may not seem the likeliest of bedfellows, experimental poetics and geography have interwoven to form a surprisingly long American tradition. Chapter 2 uncovers some of modernism's roots by examining

Whitman's "Passage to India" and demonstrating its engagement with teleological, cosmological geography popular in nineteenth-century America. This body of knowledge sought to create a master view of the entire world as one harmonious whole reflecting a divine plan, and assumed that the latest scientific methods could reveal the unfolding of that divine plan. Two of the most prominent geographers of the period helped to shape this collective project, whose work Whitman knew: Alexander von Humboldt, whose *Kosmos* was published between 1845 and 1862, and Mary Somerville, whose *Physical Geography* was first published in 1848 and subsequently republished seven times during Whitman's lifetime. Utilizing previously overlooked archival evidence, I map Whitman's cosmological oneness in relation to Humboldt as well as Somerville's work, which is far more overtly spiritual than Humboldt's and which was discussed in contemporary reviews in terms that explicitly pitted the global unity of cosmological geography against nationalist and imperialist projects. Given Whitman's influence on later modernist experimental poetics, to acknowledge Whitman's relationship with geographers like Humboldt and Somerville is to recognize that later modernist writers are genealogically entwined at their root with the knowledge formation of geography. The possibilities for global connectedness that Whitman establishes in "Passage to India" provide the foreground and the field imaginary for what twentieth-century modernist writers can do with their experimental poetics.

If Whitman pursues global oneness, the twentieth-century modernists examined in subsequent chapters reveal a varied set of projects that are more tactical and more globally circumscribed. In Hughes, Stein, and H.D., we can observe geopoetic strategies of race, nation, and empire. Hughes uses an anaphoric style of parataxis to reconstruct an African diaspora through landscape homologies; Stein's linguistic experiments with suppressing the axis of vocabulary choice measures and unmeasures national distinctiveness and American exceptionalism; and H.D.'s montage parataxis, similar to Hughes's, creates landscape transitivity that applies pressure to the categories of both nation and empire.

As I discuss in chapter 3, African American writers had to contend with the rise of environmental determinism and its implicit authorizing of colonialism and racism, which exacerbated historical traumas impacting African American relationships with the physical environment: slavery's removal of Africans from African landscapes and its implementation of forced slave labor in the global south, colonial seizure and administration of land in Africa, and lynching's use of remote wooded areas. Chapter 3 examines African American lyrics from the 1920s that revise and critique the geographical mathematics of environmental determinism, deforming the mastery of that body of disciplinary knowledge, to use Houston Baker's phrase. Langston Hughes's influential poem "The Negro Speaks of Rivers" merits special attention for reclaiming the disciplinary authority of academic as well as popular geography (Hughes was a lifelong reader of the *National Geographic Magazine*), taking charge of specific epistemological and institutional mechanisms of dominant white culture for black diasporic ends, and for using poetic form to replace American exceptionalism with an African diasporic geography.

Both Whitman and Hughes accomplish their reworkings of the globe through transportation systems: for Whitman it is both a train's-eye view and a ship's-eye view that helps to connect distant locales; for Hughes in "The Negro Speaks of Rivers" it is a river's-eye view, with an element of the train's-eye view implied by the origin story of the poem's creation. We see in chapter 4 a similar grounding in transportation as a space-traversing technology in Gertrude Stein's *Geographical History of America*, where considerations of national difference draw from the airplane's-eye view. This multigenre text published in 1936 engages directly with the genre of the geographical history generated by Semple and her cohorts even as it is also a philosophical treatise, puppet show, dramatic play, autobiography, prose poem, and a public relations plug for Stein's genius. This text demonstrates the range of formal and generic conditions in which geopoetics can operate since it uses disjunctive formal techniques similar to those employed in more recognizably "poetic" modernist texts. Stein relies upon the American exceptionalist notions that we find in texts like Semple's *American History and Its Geographic Conditions*, originally published in 1903 and revised and republished in 1933 shortly before Stein's American lecture tour and her subsequent writing of *Geographical History*. And yet Stein draws persistent comparisons between America and France and deconstructs the very categories of place and attachment she seems to evoke. The *Geographical History* thereby manufactures what I call an (anti)comparative nationalism that allows Stein to position a sense of her own American belonging against that of Franklin D. Roosevelt and to simultaneously work toward restoring a more "American" version of America to itself.

Ultimately, however, the genius of the modernist writer exceeds the categories of geography itself, or at least these particular disciplinary and disciplining moves constructed by environmental determinism. Chapter 4 concludes by exploring the similarities between Stein's poetics and the development of possibilism within the well-established French academic geography of Paul Vidal de la Blache. Taken as a whole, my analysis of Stein provides an extreme test case illustrating that experimental modernist texts might be fundamentally incompatible with nationalist projects.

H.D.'s representations of global locales in her World War II epic *Trilogy*, taken up in chapter 5, is not driven by the same transportation-based perspective adopted by Whitman, Hughes, and Stein. *Trilogy* is a purposeful dramatization of H.D.'s decision to stay in London, living just blocks away from the battery guns of Hyde Park when England was under direct attack by German forces, and of the material and metaphorical limitations British citizens were forced to accept under conditions of war. She does share Stein's strategic use of environmental determinism in relation to the category of nation; however, *Trilogy* deploys some of its core principles in order to sidestep the boundaries of nation and combines this with an environmentalist undermining of empire. Chapter 5 pays close attention to the relationship that the poet-prophet in *Trilogy* has to the war-torn London setting and her attempts to blend it with the spaces and spiritual materials of Egypt's Luxor and Karnak. Environmental determinism gives H.D., an American expatriate, the tools to be at one with the British civilian populace: to be shaped by the British environment makes the poet-speaker a part of the culture to which the environment is tethered. In what I call the logic of landscape transitivity in the poem, the speaker can also claim access to Egypt and its uplifting spiritual and mythological materials through a connection to the Egyptian landscape. This in turn produces a sort of empire vertigo that partly replicates the orientalist moves of Egyptian travel writers H.D. had read and partly registers sensitivity to Egypt's contemporary colonial status.

The book's conclusion returns to a sense of how these global structures of connection and comparison in both poetry and geography revise established ideas in modernist studies. I consider the broader implications of my geopoetic method that seeks to historicize the spatial strategies of literary texts—even experimental poetic ones—in the context of the reigning geographic epistemologies of the day. Environmental determinism as a lens can open up a range of other twentieth-century texts such as T. S. Eliot's *The Waste Land*, as can new, competing geographic paradigms that entered the academic geography scene in the mid-1920s. My conclusion briefly sketches connections between experimental poetry and these other academic paradigms: the possibilism and regionalism of Carl Sauer, which we can link to the strategies of a text like William Carlos Williams's *Paterson,* and the geopolitics of American and European geographers, a brand of thinking that stressed territorial enlargement as a geopolitical imperative, which I link to Ezra Pound's *Pisan Cantos.* In such modernist poetic texts, as in the ones I consider in detail, writers use geography and form together to generate political interventions along multiple scales of identification: for the individually authorized artist, for culture, and for larger collectivities of the nation, transnation, and the globe.

1

Academic and Popular Geography

Global Connections, Environmentalist Style

In 1920, shortly after its transformation into one of the most influential modernist "little magazines" of literature and culture, the *Dial* published an essay, "Poetry—Our First National Art," trumpeting the national value of American poetry. James Oppenheim, a political radical, antiwar pacifist, poet, playwright, and editor of another little magazine, the *Seven Arts,* declared that fiction is a lost cause because it does not convey the "essential character" of "our typical American" (Kingham 399; Oppenheim 238). This is because the writer of fiction has to "gather his expression round a typic character," but the poet can "vaguely allow that mystic depth which is common to all men." However problematic these distinctions may be, Oppenheim claims special status for the poet who, unlike the fiction writer, allows this "mystic depth" to be strained "through the sieve of the American environment, the American ideals, manners, scenery, and chaos," generating a product that possesses "what might be called the American flavour" (240). For "it is not by saying, 'Go to, I will be American,' but by allowing the direct impact of environment and the direct response, that [writers] will produce a truer and more American art" (242). Oppenheim assumes that, because American culture does not possess its own long-standing, heritable literary tradition, operations of environment play a particularly powerful role in shaping American language, culture, and art: "We have no folk, no soil song or literature: we have only our American speech, the resultant of new environment, mixture of races and new experience. This American speech is decidedly different in flavor and construction from English speech. It is not Colonial, but native, that is environmental" (240).

Such comments would have resonated with audiences who by 1920 had encountered similar assertions in a range of academic and popular sources, thus allowing Oppenheim to authorize his claim about poetry's importance for American literature and culture.[1] For Oppenheim and his readers, "environment" meant the same kind of natural, physical environment that is the concern of early twentieth-century geographers but could easily stretch to include social factors that imprint upon individual writers as well. In advancing poetry's unique importance for American literature and culture, Oppenheim broadens the terms of environmental determinism while adhering to its fundamental assumptions: that environmental pressure is a causal agent and human adaptation is its result, and that human exposure to a particular kind of environment produces similar predictable and repeatable social effects. It is fitting that environmental determinism, a theory about adaptation and conditioning, was itself adaptive and migratory in the early twentieth century as a discursive and naturalizing construct about issues of authorship, cultural identity, and community, participating in ever-broadening cultural conversations about national identity and belonging, human social expression and innovation, and human-environmental relations. "Poetry—Our First National Art," in short, suggests the uses to which geographic discourses in general and environmental determinist ideas in particular were put by the culture at large, by literary critics, and by poets in particular.

This chapter provides a genealogy of the emergence, diffusion, circulation, and aestheticization of key ideas in nineteenth- and early twentieth-century geography with which Oppenheim and the twentieth-century American modernist poets I consider here engaged. I begin by examining the end of Enlightenment geography, specifically as demonstrated in the cosmological geography of Alexander von Humboldt and Mary Somerville, in terms of its impact on American ideas about global unity and global connectedness. These concepts were important to Walt Whitman's later poetry, as I discuss in chapter 2, as well as to the global sensibilities of American modernism that followed in his wake. I then turn to the rise of environmental determinism in academic geography, which reached its apex in America in the 1910s but lingered in more popular forms of geographic writing well into the 1930s, sometimes in altered shape. While these two phases of geographic knowledge—cosmological and environmentalist—differed in their approach to the relationship between the physical earth and human society, they provided Americans with forms for conceiving of global linkages and comparisons that moved above (or below in the soil, as the case may be) the nation-state as an organizational category.

The interconnected world of cosmological geography projected an idea of harmoniousness that overrode individual regional and national identities. In early twentieth-century geography, environmental determinism made available to American writers ways to imagine geographical homologies and forms of geographical transitivity that eclipse geographical distance and cultural/national difference and that work against the salience of regional and national affiliation. That is not to say that cosmological and environmentalist ideas did not serve Americanist interests, because they did so in various ways. But recognizing their circulation in American culture reveals a more complex terrain on which nationalist and extra-nationalist sensibilities coexist, mingle, and clash. Such a historicized perspective reveals multiple and contested ways of thinking about modes of belonging that scale both upward toward the planetary and downward toward the national and that, as a result, both undercut and endorsed nationalism.

One important component of this genealogy is the way that geography becomes more specialized as a professional academic, university-based discipline at the same time that it becomes popularized in middlebrow, widely read, and more cheaply available forms. Both of these developments contributed to a larger climate in which geographic knowledge and literary modernism could comingle through the ways that they blurred or even collapsed the boundaries between the categories of the scientific and the literary. Part of what emerges across an array of academic and popular geographies is an aestheticized version of geography that operates as a strategic environmentalist style: landscape first and human society second, framed by geologic or environmental rhetoric. In the hands of modernist writers, this formula becomes a versatile politicized technique that can be used, rejected, or transformed, sometimes within the same literary text. In a sense, the genealogy of this chapter helps to explain how Oppenheim's essay in the *Dial* came to do what it does with environmentalist discourse, mapping the surprisingly long timeline along which geography and poetry have been intertwined.

European Geography in America

Traditional narratives about geographical knowledge in nineteenth-century America shift perceptibly when we consider the transnational circulation of

European and British ideas about geography. American Studies scholars and historical geographers have focused primarily on the impact in American culture of homegrown forms of cartography, geographic primers and readers, and geographic forms of entertainment. These constructed America as an imagined community in Benedict Anderson's sense and, in the latter half of the century, managed national anxieties about expansionism that continually modified America's borders at a dizzying rate.[2] For their part, historians and historical geographers have stressed the degree to which professional geography operated in its earliest stages as an extension of government work, often at the federal level and driven in part by the ways that the frontier thesis of manifest destiny and governmentality went hand in hand.[3] With Senator Thomas Hart Benton as one of the most famous mouthpieces for manifest destiny rhetoric, the mandate to explore the American western frontier required topographical mapping that would aid land settlement projects.[4] After the Civil War's disruptions, mapping the American landscape again became a preoccupation, signaled by the formation of the United States Geological Survey in 1879, which classified types of land, charted geological structures, and took note of mineral resources (Schulten 23).

These were not the only sources of geographic knowledge with which Americans came into contact, however; exclusive focus on geographic ideas produced by a single-nation framework elides the transnational circulation of geographic ideas that were not aligned with the needs of the nation-state. The nineteenth century was marked by robust transatlantic exchange of scientific ideas, with prominent theories and practices taking root on either side of the Atlantic. British and European scientists routinely engaged in lecture-tour travels in the United States (Pandora 349), and their ideas were not limited to the orbit of intellectuals and experts. Whatever impact ideas from across the Atlantic might have had in nineteenth-century American culture was magnified by the palpable shift during this period away from the practice of deferring to "expert" knowledge; in the democratized approach to science, nonexperts felt justified in wanting to try on for size the latest scientific ideas themselves. As one Wisconsin newspaper article in 1842 illustrates, the tendency was for people to not wait "for the wise men and the doctors to analyze and investigate it, before they can venture an opinion on its merits and demerits; but they take up the subject at once themselves; they reason it over in their own mind, and discuss it among their neighbors" (qtd. in Pandora 351).

In this climate of "democratized" science, the cosmological geography of two major figures in nineteenth-century thought, Prussian geographer and naturalist Alexander von Humboldt and Scottish scientist Mary Somerville,[5] found a major foothold in American culture and brought with them a way to perceive the world in global terms that operated independently of nation-state thinking. Humboldt and Somerville's approach to geography assumed that the earth and all its biological inhabitants were organized according to a divine plan that would work toward perfection, a positivist bent that some scholars have called "the old doctrine of teleology" (Livingstone 212–13) and others have called "progressive science" (Reynolds 242). This approach to geography taken by Humboldt, Somerville, and other European figures such as Karl Ritter assumed that geographical data concerning the configurations of the continents, the differentiation of distinct areas, and the distribution of plants, animals, and humans all pointed, in a Kantian way, to the existence of a great spirit architect whose plan unfolds on the earth and that human society was being pushed along toward a state of perfection (G. Martin, *All Possible* 123, 138). These European traditions of geography in the nineteenth century marked the apex of the model of "earth description," what Alexander von Humboldt called *Erdbeschreibung,* which attempted complete synthesis of knowledge about the entire world (G. Martin, *All Possible* 120), the comprehensive nature of which was enabled by an integrative model of science that drew upon astronomy, physics of the earth, biology, geology, and human distribution and behavior (Neeley 133). It is the worldness, if you will, of Humboldt's and Somerville's projects that I suggest needs to be considered as an important organizing principle that nineteenth-century Americans could utilize in order to consider geographical and spatial forms of belonging that existed alongside but were not eclipsed by the many homegrown forms of geographic knowledge that supported nationalist projects.[6]

The impact and reach of both Humboldt and Somerville in the nineteenth century is impressive. Humboldt most famously articulated his ideas in his multi-volume opus *Kosmos,* published between 1845 and 1850 and immediately translated into English as *Cosmos: A Sketch of a Physical Description of the Universe.* By 1851, eighty thousand copies of the book had been sold (Reynolds 244). Emerson wrote that the nineteenth century was "the Age of Humboldt," reflecting the ways that Humboldt was, as one scholar has put it, the nineteenth-century equivalent of Albert Einstein.[7] Humboldt's *Cosmos* was certainly one of the more prominent sources for the way Americans thought about science and culture (Pandora 352). It is no surprise, then, that Humboldt's iconic status has overshadowed the influential work of his contemporary scientist and friend Mary Somerville.[8] But Somerville's contributions were not just well regarded by Humboldt himself, who praised her work (Neeley 230); they also found a receptive, eager audience in Europe and the United States. Her own "earth description" geography book, *Physical Geography,* first appeared in 1848 and became widely used as a university textbook throughout the rest of the nineteenth century. Its constant revision and republication helped to extend the temporal range of the circulation of cosmological geography in key ways. Somerville revised and republished *Physical Geography* six times (1848, 1849, 1851, 1858, 1862, 1870) (J. Baker 209n3), with a final edition published after her death in 1877 by John Murray and one of Somerville's daughters. By contrast, Humboldt produced each of the subsequent volumes of *Cosmos* between 1845 and 1858, with the final volume appearing just after his death in 1859 (Kellner 199). Official estimates that *Physical Geography* sold approximately sixteen thousand copies (Creese 203) pale when compared with Humboldt's figures; however, whatever estimates we have for Somerville's work are surely conservative given the ways that her work was repeatedly reprinted without permission by American publishers in an age before transatlantic copyright laws. Somerville complained that her translation of Laplace's *Mecanique celeste* (Celestial Mechanics) was pirated and illegally published in America by Cary and Lea of Philadelphia in 1831; her second book, *On the Connexion of the Physical Sciences,* was pirated and published in the United States at some point in 1834 as well (Creese 202–3). The practice of reprinting European books waxed and waned in American publishing throughout the nineteenth century but remained a common practice until the 1880s.[9] So it comes as no surprise that various editions of *Physical Geography* appeared in America in pirated form as well.[10]

Americans were receptive to her ideas. George Perkins Marsh, for example, "found her observations about man's destructive use of the earth very stimulating" (qtd. in Dikshit 89; also see Livingstone 272–74). Emerson implicitly praised Humboldt as a scientific visionary who taught the principles of unity (Walls, *Emerson's Life in Science* 217). But he, like many reviewers of the period, explicitly compared Humboldt and Somerville in terms favorable to both (Walls, *Passage* 362n11). Stand-alone reviews in American periodicals praised Somerville as well; she received high praise for each of her early editions of *Physical Geography* in one of the most popular nineteenth-century American periodicals, *Godey's Lady's Book.* Prominent reviews in British periodicals like the *North British Review* were almost immediately reprinted in American periodicals such as the *Eclectic Magazine of Foreign Literature,* though key figures in the American Renaissance like Whitman also routinely read the *North Brit-*

ish Review itself.[11] In fact, Whitman kept a copy of the *North British Review*'s lengthy praise of *Physical Geography* in his files, commentary that likely encouraged the Somervillian influences that chapter 2 traces in his work.

The notion of an ordered, harmonious universe celebrated in Humboldt's and Somerville's work promoted an idea of global interconnection made possible by the divine, a model that deemphasized if not outright ignored human-generated artificial geographical divisions (Neeley 126). Unlike popular geography readers of the late eighteenth and early nineteenth century such as Jedidiah Morse's *Geography Made Easy*, which reflects issues of concern to national identity,[12] these transatlantic imports adopted a vastly broader aperture, stressing planetary forms of physical, biological, and human interconnectedness. Humboldt's *Cosmos*, for example, describes physical geography as involving the "contemplation of all created things, which are linked together, and form one whole, animated by internal forces" (qtd. in Kish 410). His emphasis lies on tracing the planetary distribution of living things and their interdependent relationships, viewing the entire world as one big connected, integrative system:

> In the exposition of the terrestrial portion of the *Cosmos*, it will occasionally be necessary to descend to very special facts; but this will only be in order to recall the connection existing between the actual distribution of organic beings over the globe, and the laws of the ideal classification by natural families, analogy of internal organization, and progressive evolution. (Humboldt 1: 43)

Humboldt admits that the divisions of countries are essential, as they supply us with "the most important materials for the composition of a physical geography," but he avers that listing descriptions by country

> would as little give us a true image of the general conformation of the irregular surface of our globe, as a succession of all the floras of different regions would constitute that which I designate as *Geography of Plants*. It is by subjecting isolated observations to the process of thought, and by combining and comparing them, that we are enabled to discover the relations existing in common between the climatic distribution of beings and the individuality of organic forms. (1: 42)

Physical Geography adopts this approach, although Somerville makes explicit the ways this sense of the global overrides or elides the salience of the nation as an epistemological category:

> *Physical Geography* is a description of the earth, the sea, and the air, with their inhabitants animal and vegetable, of the distribution of organized beings, and the causes of their distribution. Political and arbitrary divisions are disregarded, the sea and the land are considered only with respect to those great features that have been stamped upon them by the hand of the Almighty. (1)

For Somerville, human territorial markers or divisions matter less precisely because they do not stem directly from a divine creator; instead, she favors features of the earth and its biosystems that can reveal a divine plan at work. Not only do these texts foreground scales of belonging that exceed the nation-state, but they also adopt a comparativist methodology that encourages a centrifugal geographical outlook. Humboldt rejects quantifying qualities of geographic science and instead stresses the utility of generalizable, impressionistic comparisons that encourage this kind of outward-oriented perspective: "It is only the general results of comparative orography [the study of mountain ranges] and hydrography [the study of the earth's water systems] that belong to the science whose true limits I am desirous of determining, and not the special enumeration of the greatest elevations of our globe of active volcanoes, of rivers, and the number of their tributaries" (qtd. in Kish 409). While these kinds of projects in cosmological geography did utilize direct observation of dynamics of particular regions, especially in the case of Humboldt who drew upon his own extensive travels, their method depended heavily upon generalizations borne of comparison and analogy (Neeley 135).

These texts, significantly, performed such global, extra-national sensibilities on a formal level. In stark contrast to American-produced geography primers that portrayed American geography in relation to specific countries in other parts of the world, or to much of the other geographical writing of the period that described major geographical features country by country, the very organization of Humboldt's and Somerville's texts finds little use in national markers. They both arranged material by physical category or environmental features (geology, astronomy, water systems, etc.) in order to grasp the underlying factors responsible for the interconnections among various phenomena in nature and for the nature of the placement and distribution of plant and animal life, presented as part of one unified picture (Creese 203). When their texts do aggregate downwards in terms of geographical scale, they only begin to approach questions of national concern in addressing the formation and character of the various continents and only implicitly in their limited discussion of the nature of racial categories.

Although cosmological geography exerted its biggest and most immediate impact in the first half of the nineteenth century, its legacy was an enduring one, especially for writers of the American Renaissance such as Whitman, who encountered these ideas from multiple sources and revisited them across the long span of his career. Somerville's own reputation may have suffered by comparison in historical assessments of the field because her *Physical Geography* was initially published—and continued to be reissued—after Humboldt's *Cosmos* first appeared, but the very fact of that delay extended the period of injection of cosmological geography into the scientific community and into broader cultural discussions about science and the globe. The numerous reissued editions of *Physical Geography,* combined with the professional accolades Somerville won late in her career, such as the Royal Geographic Society Medal in 1869, helped to further the circulation of cosmological geography well into the nineteenth century.

Environmental Determinism: Global Comparison, Environmentalist Literary Style

The next major phase of geography, environmental determinism, shifts radically away from the positivist, romantic qualities of cosmological geography in a number of key respects. But environmental determinism continues cosmological geography's outward-facing spatial orientation through the global comparisons it needs to make in order to generate claims about the ways environment shapes culture. These comparativist gestures similarly undermine the power of the nation-state to organize ideas of identification and belonging, even if they serve a higher purpose for the academic geographers or the political or governmental bodies who apply them to nationalist or imperialist causes.

As the nineteenth century came to a close, geography in the United States became a more narrowly drawn discipline and a more academically rooted profession at roughly the same moment that literary modernism was starting to gain momentum. This transformation within academic geography was enabled in central ways by environmental determinism, the next major paradigm, which took root in the field in the 1890s, blossomed in the 1910s, and lingered well into the 1930s in new forms. Its central premise—that the natural environment operates as a set of controls or conditions to which human society responds—tightened for Americans the perceived connection between geographical environment and culture and brought into being cultural associations that could treat the former as an index to the latter. This paradigm supported American nationalist and imperialist projects because it allowed for differential distinctions between more "civilized" societies—advanced because gifts provided by the environment allowed them to modify or even conquer it—and primitive societies in which these gifts were lacking. As scholars have demonstrated, environmental determinism thus gave geography legitimacy by generating geographic knowledge capable of rationalizing the imperialist practices of western nations.[13] But environmental determinism also brought with it several new ways to think about the relationship between the natural environment and human society and how to orient oneself in relation to the globe. Even while environmental determinism upheld ideas of western superiority as evidence of the optimal development of the natural advantages that geography presents, it also provided the formal tools of transnational comparison useful for modernist projects. Environmental determinism in this sense marks a revival of the global, comparative orientation of the nineteenth-century cosmological geography of Humboldt and Somerville, albeit with a different shape and in a way that radically reduces the amount of agency given to the human side of the environment-human equation.

Recognizing the internal, comparative logic within environmental determinism means recognizing that, like many discourses, it contains multiple and sometimes contradictory ideologies: although environmental determinism's main compass is western superiority or American exceptionalism, the logic of comparison and environmental homology it employs effaces the salience of the nation-state and the civilizational distinctions that would set the West apart from "the rest." As this tenacious discourse travels further from its academic geographic origins, it becomes an increasingly aestheticized, flexible form for American culture and for modernist writers.

In its initial institutionalized inception in the United States, academic geography was most concerned with establishing its own legitimacy as a field of inquiry, and its operations in this early sense did nothing to decrease the distance between geographic knowledge and literary production. Environmental determinism emerged at a moment when the sciences were becoming increasingly specialized. As the paradigm of cosmological geography in the nineteenth century illustrates, geography had long enjoyed a catholic, far-ranging scope that included geology, botany, physics, astronomy, geometry, history, and the like, to the extent that it was considered "propaedeutic" in Kantian terms, a foundational discipline upon which more advanced study of other things would rest (Koelsch 246). Concerned that the perception of geography as the "mother of all sciences" would tarnish its legitimacy by being seen as both everywhere and nowhere in particular, British geographer Halford J. Mackinder warned of the ways that a discipline that drew upon such disparate fields would not fare well in this kind of climate.[14] If geography were to include the human element, albeit only when it could be shown to be a result of geographic influence, the field could differentiate itself from physical sciences such as geology and botany and social sciences such as sociology (Koelsch 252). In providing geography with its more specialized form, environmental determinism became a large part of its institutionalization in American universities. Although geographers took multiple approaches, there was a resounding sense of uniformity in the discipline regarding the basic importance of geographical influence upon human society. By 1914, a survey of academic geographers ranked the studying of environmental influence as the biggest feature defining geography of the time (G. Martin, *All Possible* 375).[15]

A whole generation of environmentalist geographers contributed to this central paradigm shift, and most of them studied directly with William Morris Davis. Routinely referred to as America's most prominent first-generation academic geographer, Davis's conception of *ontography* brought together *ontogeny* (the nature of the development of individual beings) with *ontology* (the nature of being) in the study of the impact of the physical landscape on human society.[16] With this approach, Davis differentiated geography from fields like geology because of its consideration of human elements, initiating what David Livingstone has called "the geographical experiment," which was "a maneuver designed to hold together the natural and social worlds under one explanatory umbrella" (Livingstone 210). His idea that environmental conditions initiated a human response also hewed more closely to "scientific" models of cause and effect that would lend the field credibility in an era of increasing specialization (G. Martin *All Possible* 347; Schulten 75). Several of his students rose to prominence in developing an array of environmentalist approaches, the best known among them Albert Perry Brigham, who took an environmentalist approach to region, and Ellsworth Huntington, who developed his own brand of determinism in relation to climate. Although the field was marked by the subtle differences among these major practitioners, their approaches to the relationship between the natural environment and the human are all indebted to varying degrees to Lamarckian ideas about adaptation to the environment through acquiring and passing on advantageous characteristics.[17]

Ellen Churchill Semple operated in this environmentalist, neo-Lamarckian vein as well, although her work did the most to generate a form of environmental determinism that was not only hospitable to popular readers but also invested in forms of interdisciplinary exchange with history, sociology, and literature in particular. Although she no doubt benefited from Davis's influence, she studied directly with German geographer Friedrich Ratzel, who advanced in his two-volume *Anthropogeographie* (1882, 1891) a fairly rigid, deterministic set of ideas about the natural environment as a set of social controls. Semple adapted but modified Ratzel's ideas by asserting not a deterministic dynamic per se but a focus on the "geographic factors and influences" that impact human society (*Influences* vii).[18] The most rigid formulation Semple offers appears in the very first sentence of *Influences*: "Man is a product of the earth's surface." But the lyrical language she employs almost immediately gives this relationship a maternal, nurturing sort of cast: "the earth has mothered him, fed him, set him tasks, directed his thoughts, confronted him with difficulties that have strengthened his body and sharpened his wits, given him his problems of navigation or irrigation, and at the same time whispered hints for their solution. She has entered into his bone and tissue, his mind and soul" (1). Semple's language of umbilical connection is typical of the lyrical qualities of the volume, although the environment-human connection is most often expressed in terms of the idea of "influence" that appears in her book's title: "Man has been so noisy about the way he has 'conquered Nature,' and Nature has been so silent in her persistent influence over man, that the geographic factor in the equation of human development has been overlooked" (2). Indeed, as I will discuss below, Semple's focus seems less on arguing for geography's dominion over the human and more on advancing an early version of the "spatial turn" that we now see animating much contemporary work in the social sciences and the humanities. She offers eloquent justification of geography's relevance to modern human society and its resulting importance as an epistemological project in the broadest intellectual sense. "Every clan, tribe, or state or nation includes two ideas, a people and its land, the first unthinkable without the other" (51).

Semple has been most visibly associated with this paradigm of geographic thought because her scholarship singularly pursued this topic for more than two decades. I would add, however, that her work also did the most to facilitate the spread of environmental determinism in American culture and to enable the conditions for its usability by American writers. In an academic context, Semple enjoyed singular success as one of the very first female professors at Clark University, then home to one of the strongest new departments in academic geography, and as a decorated and often-requested lecturer at universities and academic and popular science and history societies. Her 1903 *American History and Its Geographic Conditions,* which, as Innes Keighren notes, mainstream publisher Houghton Mifflin had engaged her to write (33), applied environmentalist principles adapted from Ratzel to American history and culture. It was reviewed well by geographers and also by Frederick Jackson Turner, who wrote about it in favorable terms for the *Journal of Geography* (Keighren 34). Its broader cultural impact can be gleaned from the speed at which the book was quickly adopted as a standard textbook on both environmentalist geography and historical geography; not only was it placed on reading lists for geography and history at multiple universities, but copies were placed on every ship in the U.S. Navy, and it was also among the required reading of students at West Point (Keighren 33-34). Well into the 1930s, scholars still considered this book to be a standard reference for anyone interested in American history and geography.[19]

Public visibility almost immediately followed the publication of Semple's second and most influential book, her 1911 *Influences of Geographic Environment.* Upon her death in 1932, the *Geographic Review* (the academic journal associated with the American Association for Geographers that Davis founded) perhaps predictably called her "America's most outstanding geographer," citing her *Influences* "as the most scholarly presentation in English on the doctrine of the influence of geographic environment" (Whitbeck 500, 501). But what may be surprising is that Semple promoted the book on an eighteen-month lecture tour around the world and that over forty periodicals reviewed the book, ranging from academic humanities journals like the *Yale Review* to mainstream magazines like the *Nation* to everyday newspapers like the New York *Post-Standard* and *Sun* (Keighren 51–53). Some of the most compelling evidence for the broad-ranging impact of Semple's work lies perhaps in the fact that African American periodicals took notice of her work. For example, in its November 1911 issue, the *Crisis* includes *Influences* in its list of notable books, just a few lines below W.E.B. Du Bois's *Quest of the Silver Fleece.*

This is not to say that all reviews were positive. In fact, some of the overtly negative responses to Semple's work voiced an alternative set of investments in the individual agency of the artist and the leader, agency that the modernist writers in chapters 3 through 5 exert when they engage with environmentalist principles but bend them to their own projects of rewriting conventional geographic space. A review of *Influences* published in an earlier iteration of the *Dial,* when it was a Chicago-based national magazine focused on the sciences, the arts, and politics and before it became the modernist little magazine in which Oppenheim's essay above appeared (Britzolakis 87), praised the usefulness of Semple's book for biologists and historians but took her work to task for not crediting human innovation and creativity: "the potency of great leaders and of the influence of ideas are powerful agencies in the control of human action" ("Environment as a Force" 399). Part of the review's objection stems from tensions across multiple operations of scale, the ways that Semple's macro-scale analysis of the regional or national societal unit fails to account for what happens on the micro-scale of the individual actor. Semple's treatment of "location, area, boundaries" and "seas, rivers, peninsulas, and islands" and the like are "potent factors in shaping the course and limiting or impelling the social and ethnic career of peoples and nations," but they offer "insufficient recognition" when they aggregate downward to the contributions of "great leaders" or those who advance innovative ideas, whose actions can operate as a "counterfoil" to the impact of environmental factors. In entering the public conversation about the relationship between the environment and society that Semple's book inspires, the *Dial* review demonstrates the ways that the arts actively recognized the contributions of environmental determinism while positing the unique innovations of the individual artist or leader. Innes Keighren's recent monograph on Semple has contextualized the impact of her first two books within academic geography and in relation to a broader periodical-reviewing public in valuable ways, but her work played a role in a much broader set of negotiations about the nature of the human-environment relationship, the way the natural environment could be conceived and represented, and the ways that Americans could orient themselves in relation to global geography.

Implicit in this debate is the issue of authorial agency itself: while Semple, or the environmentalist geographer in general, grants more agency to the environmental rather than to the social body or even the individual, it is still within the purview of the geographer to convey such social effects. When modernist writers use some of Semple's ideas and modify them for their own purposes, they are not only rescaling the macro-level patterns of environmental determinism to the orbit of the individual voice and the individual poet's vision but they are also appropriating the power of that disciplinary perspective and of representation generally. Chapter 3 demonstrates this at work most clearly when Langston Hughes draws upon environmentalist equations between environment and civilizational achievements in "The Negro Speaks of Rivers" and subordinates them to the authoritative position not of the geographer, whose typical purview it is to "speak of rivers," but of the knowing African diasporic subject. If this then begs the question of whether all writers writing about the natural environment are in some sense always already in a contest for representational authority with geographers, the stakes for authorial power and agency are most stark with environmental determinist geography, which expressly downplays the possibilities for agency in the way it portrays human society, with whatever agency remaining in the equation implicitly granted to the geographer in a way similar to that of the invisible "Seeing Man" natural history travel writers that Mary Louise Pratt describes in *Imperial Eyes.*

Part of the work involved in tracing the larger cultural debates about how the environment and the human can be conceived hinges upon recognizing Semple's multiple efforts to engage with a nonacademic public. All three of her books were written to engage a broad audience in ways her reviewers praised. By contrast, fellow geographic environmentalist Ellsworth Huntington's *Civilization and Climate* (1915) appealed more to expert than to nonexpert readers. Huntington's work was as well known to scholars and reviewers as Semple's at the time; however, in its attempt to measure various instances of environmental influence numerically, his approach to environmentalism was more quantitatively driven than hers. Although Huntington dismissed Semple's work as being too "impressionistic" (G. Martin, Review 133), it is precisely this impressionistic quality that allowed Semple's work to bridge the distance between geographical science and other cultural fields, including the realm of the literary.

Its usability and its ability to provoke went hand in hand in a number of other respects. The comprehensive approach Semple took to geographical location provided a wide platform on which other geographical and literary writers could build. Compared to her peers, Semple approached the environment in the most inclusive way, considering topography, climate, botany, agriculture, economic structures, population distribution, culture, politics, and literature. Unlike Huntington's *Climate and Civilization,* which, as his title suggests, dealt primarily with climate, Semple's ideas about environment were all-encompassing; location, for her, meant "climate and plant life at one end of the scale, civilization and political status at the other," and it also involved the ways a particular location related to surrounding territories: "the influences of a land upon its people spring not only from the physical features of the land itself, but also from a wide circle of lands into which it has been grouped by virtue of its location" (*Influences* 131).

Part of Semple's appeal to a constituency beyond the boundaries of academic geography stems from her own advocacy of geography and space, her positioning her field as a valuable missing epistemological ingredient in other disciplinary formations, if not as a superior kind of project from which other fields should borrow. Semple's efforts to make others reckon with the significance of the geographic environment anticipate the contemporary "spatial turn" of recent years which has called for a new examination of the spatial dimensions of social life. The sort of assertion Edward Soja makes in *Postmodern Geographies,* discussed in the introduction, that a long-standing historicism "actively submerges and peripheralizes the geographical or spatial imagination" (15) resonates with Semple's own arguments about the overprivileging of history: "all these sciences, together with history so far as history undertakes to explain the causes of events, fail to reach a satisfactory solution of their problems largely because the geographic factor which enters into them all has not been thoroughly analyzed" (*Influences* 2). In advocating for the significance of geography, she turns to Kant's question, "Which was there first, geography or history?" and his answer, "Geography lies at the basis of history." This line of argument leads to Semple's claim that "all historical development takes place on the earth's surface, and therefore is more or less molded by its geographic setting" (10). It is not that the importance of history is denied, but rather that the emphasis needs to be redistributed, with geography as the central mover of historical development and hence of history itself. The geographer "therefore regards history in no small part as a succession of geographical factors embodied in events" (11). For these reasons, the environmentalist geographer with a historical bent is superior to the mere historian; she is "uniquely poised to understand historical evolution in ways the historian might miss, since the faster and more favorable the geographic base of a society, the greater may be its ultimate historical significance" (53).

Semple makes a similar claim about sociology's anemia compared to all that geography can grasp. "Most systems of sociology treat man as if he were in some way detached from the earth's surface; they ignore the land basis of society," but the environmentalist geographer is especially suited to recognize factors important to the sociologist like economic and psychological structures, as well as the ways that modern society has "grown into every foot of its own soil, exploited its every geographic advantage, utilized its geographic location to enrich itself by international trade, and when possible, to absorb outlying territories by means of colonies" (*Influences* 53). Semple, in short, advocates broadly for geography's centrality in a way that is made possible by the multiple boundaries she crosses that her peers did not: the distinction between academic and nonexpert readers, between scientific and literary style, and between geography and other disciplines.[20]

Semple's exhortations about the necessary geographical reorientation of knowledge production extend directly into the realm of the literary. Semple conscripts the disciplinary category of literature most overtly as a key terrain through which she can demonstrate geography's centrality in other disciplinary and aesthetic forms. This colonization, or even cannibalization, of the literary informs the extensive "Literary Reading Lists" appendix Semple added to her new edition of *American History and Its Geographic Conditions* (*AHGC*), revisions of which she worked on until her death in 1932 and which was published posthumously in 1933. This edition develops or clarifies some of the book's historical points and, through the aid of her Clark University colleague Clarence Fielden Jones, a professor of economic geography, includes new economic material. However, the creation of a new appendix containing both a "Supplemental Readings" section of added geographic sources and the "Literary Reading Lists" section containing more than 240 briefly annotated items is the most obvious change, although it is one that has somehow completely escaped the notice of historians and historical geographers. What it conveys to a broader American readership is a fundamental way to imagine identity and culture through the influence of landscape, rather than treating landscape as a passive container or backdrop, and in the process instantiates a usable environmental determinist canon that includes American literature, especially poetry.

Semple largely organizes this appendix by geographical region within and around the United States rather than by discipline or literary genre, suggesting that particular literary features are less important than the impact of the geographical conditions in which authors write or in which their literary texts are set. The cluster of literary texts focused on Alaska, North Arctic America, and Canada, for example, "shows the adjustment of life to local geographic conditions" and identifies specific ways that the population has adjusted (*AHGC* 469). Throughout this section of the appendix, biographies and historical volumes about particular regions are occasionally mixed in without distinction, and examples of realism such as Melville's *Moby Dick* appear alongside examples of naturalism or of regionalist texts such as Sarah Orne Jewett's *Countries of the Pointed Firs* simply because they all reflect the influence of New England as a region. Semple explains that her purpose with the "Literary Reading Lists" is to "reveal the rich material often contained in general literature coming from writers familiar with certain regions where they have lived and worked; to encourage an appreciation of a choice literary style, even in dealing with scientific facts; and finally to feed the imagination of youth, rather than restrict it to the starvation diet which constitutes the major part of American education" (469). We can easily imagine that urging readers to gain "an appreciation of a choice literary style, even in dealing with scientific facts," applies equally to Semple's own "impressionistic" literary approach to geography, to cite Ellsworth Huntington, and to the literature she enlists for her cause.

These forms of authorization operate on another level as well: even while constructing an environmentalist canon of literature into which she places these texts, Semple confirms the robustness of her paradigm by linking it to the dynamic field of American-literature-in-the-making. Although the appendix is "merely typical and suggestive," Semple asserts very plainly that these literary reading lists "assume special significance in view of the enormous amount of literature being produced in American history, both academic and fictional" (*AHGC* vi). In positioning literature as an important way Americans could not just absorb environmentalist ideas but recognize them as such retroactively, Semple constructs a dynamic between geographical forms of thinking and literature that operated in a more complicated multidirectional interdisciplinary fashion than historiographical accounts of geography might suggest. It is telling that when Semple provides brief synopses of the texts in her appendix, she mixes up the scientific and the literary, shuttling among annotating items as "a story of," "a description of," and "a study of." Indeed, though this is called a literary appendix, which contains at most a smattering of histories and biographies, Semple includes her own study, "The Anglo-Saxons of the Kentucky Mountains" (481), a breakthrough article that she had originally published in the academic *Geographical Journal* in 1901.

Semple's "Literary Reading Lists" creates an environmentalist poetry canon that is confined to representing domestic U.S. geography in ways that many of the modernist writers discussed in subsequent chapters complicate. "Literary Reading Lists" includes an entire subsection devoted to "Poems with a Geographical Background." Its placement here reminds us of the higher visibility and status of poetry during the modernist period, and the inclusion performs the sorts of linkages across science, the humanities, and literature, even poetry, that existed during the period and that are less visible in our contemporary moment. It also highlights two parallel sets of tensions: (1) that between the nation's efforts during this time to build a distinctly American literature, as Oppenheim's comments at the beginning of this chapter attest, and the global engagements of American modernist poets such as later Whitman, Hughes, Stein, and H.D., and (2) the pursuit in Semple's *American History and Its Geographic Conditions* of a decidedly Americanist vision while other elements of environmental determinism encouraged global, comparative forms of geographic analogy, homology, and transitivity.

We can in fact observe Semple carving out a nationally circumscribed canon of American poetry from more geographically and culturally diverse raw materials. She includes, for instance, the modernist poetry of Amy Lowell, highlighting her nature-oriented imagist lyrics from the collections *What's O'Clock?* and *A Dome of Many Coloured Glass*. But in a move that is not surprising given the lumping qualities of the fiction entries in the appendix, Lowell is identified as a "New England" poet, listed alongside stylistically diverse figures such as William Cullen Bryant, Ralph Waldo Emerson, Robert Frost, Henry Wadsworth Longfellow, and James Russell Lowell (Amy Lowell's cousin, the well-known Fireside poet). This method of organization domesticates the international dimensions of Amy Lowell's involvement in the imagist movement, necessarily leaving out poems such as "The Cyclists," whose natural imagery stages visions of a dying England, or "Patterns," which deals with World War I in a European setting, in favor of poems like "Lilacs" that overtly describe the New England landscape. The purpose and overall design of Semple's poetry selections pushes to the outside of an environmentalist canon of literature twentieth-century modernist poets like Hughes, Stein, and H.D., who engage with non-American locales. This gesture is replicated, not surprisingly, with Semple's selections of Whitman. "Song of Myself" makes an appearance along with an assortment of other nature lyrics from his earlier editions of *Leaves of Grass*, which Semple places under the subheading "General Distribution" in such a way that Whitman, ever the seeker of vast audiences and the means to overcome U.S. sectionalism, would have approved. But in ways that parallel much traditional Whitman scholarship, Semple's choices elide Whitman's later poetry, which is equally grounded in the natural world but which seeks to embrace the "terraqueous globe." *American History and Its Geographic Influences* may conceive of poetry as contained by an American geographical environment, but a range of examples of modernist poetry suggests a diverse set of engagements that produce something global rather than purely national.

Furthermore, while much modernist poetry offers external challenges to Semple's national applications of environmental determinism, even within her own body of work her Americanist purposes run counter to or are complicated by some of the global, comparativist moves that she advances elsewhere. Indeed, Semple's work brings to American readers a comparativist way of thinking, making available formal structures for linking spatially distant locales. This shift gains momentum in part from the way that she downplays the role of race and ideas of essentialized racial otherness that it encoded, compared to the attention given to biologized ideas of race in her mentor Ratzel's work. By stressing environmental factors instead, Semple gave Americans ways to explore ideas of difference and belonging that were not routed through biological bodies and that opened up avenues for considering identity, affiliation, and belonging in dynamic rather than static terms. Semple made explicitly clear that she aimed to indigenize Ratzel's ideas for a contemporary and specifically American readership, explaining that her 1911 *Influences of Geographic Environment* was "adapted to the Anglo-Celtic and especially the Anglo-American mind" (*Influences* v). Semple herself seems to have given her text over to the determining influence of an American environment that was at least in theory democratic and egalitarian. She explains that "the race factor . . . was eliminated for certain large classes of social and historical phenomena" in her analysis, preferring to view human society as a constructed entity, at least insofar as cultural qualities were forged in the crucible of environmental influence. This shift brings into clearer focus the ways that literary texts such as Hughes's "The Negro Speaks of Rivers," Stein's *The Geographical History of America*, and H.D.'s *Trilogy* engage with forms of difference—racial, cultural, national—routed indexically through landscapes and locations in a manner that might otherwise not register politically in the same ways that attention to the racialized body might. And since in Semple's hands the relationship between landscape and culture is an evolving, dynamic one, it also offers modernist writers ideas of racial, cultural, and national difference as malleable, changeable concepts. That is not to say that Semple's work was free of hierarchical thinking or that it was not complicit in supporting imperialism and American exceptionalist thinking; when measuring western societies against nonwestern cultures, she repeatedly privileges Europe and America in ways that are indebted to Herbert Spencer, the figure often associated with the advent of Social Darwinism (*Influences* 59).[21] Much of her work assumes the superiority of western civilization precisely because it has successfully built upon the geographical advantages of its soil, whereas for less "advanced" cultures, "the looser is the connection between land and people . . . the lower the type of social organization" (58). But the methods and forms for thinking about the physical environment as an articulatory space for ideas of identity, belonging, and difference that Semple brings into being need not be reduced to the particular conclusions she draws, and all of environmental determinism and its apertures should not be equated with American exceptionalism and western imperialism.

In significant ways, the discourse of environmental determinism has embedded within it structures and logics of global comparison, global homologies, and ideas of landscape and cultural transitivity that support American ways of thinking globally and that align with the global projects of the American poets considered here. Indeed, Semple's work helps to codify a form through which global comparisons can and should be made based on geographic environments, aside from or in a way unrelated to national boundaries or ideas of national statecraft: "The writer's own method of research has been to compare typical peoples of all races and all stages of cultural development, living under similar geographic conditions. If these peoples of different ethnic stocks but similar environments manifested similar or related social, economic, or historical development, it was reasonable to infer that such similarities were due to environment and not to race" (*Influences* vii). At various junctures, Semple emphasizes the global scope of her project and by extension conceives of the will to know in global terms. She asserts that comprehending the evolution of human activities such as navigation, trade, and agriculture, as well as "theories of human population," "can never reach their correct and final statement, unless the data for the conclusions are drawn from every part of the world, and each fact interpreted in the light of the local conditions whence it sprung" (51).

There are, of course, several ways that Semple recognizes the organizing power of the nation-state, but notably she also sees beyond it. In a section of *Influences* devoted to "Society and State in Relation to Land," Semple tries to counter the assumption that land matters more to the nation-state than it does for social and cultural values because of the ways that wars are fought over territory. Instead, she argues that land is just as "necessary and potent" in the evolution of society because "society is far more deeply rooted in the land than is a state, [and] does not expand or contract its area so readily" (*Influences* 52). So when we consider British ascendancy over New France in the Americas in 1763 and America's expulsion of Spain from Cuba and the Philippines in 1898, "the race stocks, languages, customs, and institutions of both France and Spain remained after the flags had departed" (52). Changes in political borders due to conflict are less material than the relationship between a society and its land.

Landscape matters to Semple in ways that often override the nation-state and that, for American readers and writers, suggest routes for thinking beyond or below the nation-state as a way of defining cultural identity and belonging.

Popular Geographies, Environmentalist Styles

Ideas of environmental influence, the blurring of boundaries between the scientific and the literary, landscape primacy, and global comparativeness did not emanate in American culture from academic geography alone. These ideas took root, evolved, and spread in popular periodicals on both sides of the Atlantic, creating multiple points of contact for American modernists at home and abroad. On March 27, 1914, Semple gave the lecture "Problems of the Japanese Farmers" to the National Geographic Society in the wake of the success of *Influences of Geographic Environment* (Keighren 113). Though this lecture never appeared in the Society's signature periodical, *National Geographic Magazine,* her ideas and those of her fellow environmentalists inform the magazine's approach to world geography in numerous ways. One of the most popular "middlebrow" magazines in the United States in the twentieth century,[22] with circulation in excess of five hundred thousand by 1918 (Pauly 517), *National Geographic Magazine* was a publicly recognized marker of class and educational status because of its blend of edification and entertainment, often passed off as a marker of "good taste" and treated like a "standard-bearer" of American values. As Tamar Rothenberg has observed, "For much of the first half of the twentieth century, it was *National Geographic Magazine* that provided Americans—middle-class white Americans, in particular—with an ongoing narrative of the world and how to see it" (23).[23] A 1915 pamphlet, *The Story of the Geographic* by John Oliver La Gorce, announced the magazine's mission to help disseminate key geographic ideas to a broader public: "the Geographic Magazine removes the padlock of technical terms from the portals of geographic science and invites the world to share its delights" (qtd. in Rothenberg 23). *National Geographic Magazine* was thus poised early on to disseminate ideas from academic geography to a very large public, ideas the magazine and its imitators refashioned in their own image. In this sense, *National Geographic Magazine* extends the work Semple began in linking the scientific and the literary, even as it gradually transformed the academic mode into a distinctly identifiable environmentalist narrative style and aesthetic.

Like many Americans, numerous modernist writers were regular, devoted readers of the *National Geographic Magazine.* Scholarly accounts of modernism have largely overlooked the connections between literary modernism and this periodical, perhaps because a middlebrow magazine like *National Geographic Magazine* does not easily fit the discussions of periodical culture in modernism that were for so long dominated by the impact of the "little magazine."[24] But modernist writers contributed to the *National Geographic Magazine* and were among its most devoted readers. For instance, Joseph Conrad's popular essay "Geography and Some Explorers" appeared in *National Geographic Magazine* in 1924, before it was republished in several American newspapers and his *Late Essays*. Janet Flanner and Solita Solano contributed with even more frequency; both toured Europe and the Middle East for the *National Geographic Magazine* just before they arrived in Paris in the fall of 1922 (Fitch 135).[25] Even those modernists who did not contribute to the magazine knew of it and read it with regularity, from H.D. and Bryher to Langston Hughes. For Marianne Moore, for example, the *National Geographic Magazine* took its place alongside fragments of archaeology, Chinese philosophy, and canonical literature (Shakespeare, for example) that inform her work (Willis 3).[26] Mabel Dodge Luhan, a major patron of experimental modernist art and a columnist for the Hearst news organization, subscribed to the magazine and participated in National Geographic Society events.[27] English poet Ralph Hodgson collected National Geographic Society maps.[28] In a November 26, 1934, letter to Louis Zukovsky, William Carlos Williams describes longing for the Caribbean islands he sees covered in a recent issue of the magazine (Ahearn 208), and he explicitly describes or misdescribes a *National Geographic* picture of an African chief and his multiple wives in book 1 of *Paterson* (13). Langston Hughes, in thanking *National Geographic* editor Peter White in the 1960s for sending him an issue of the magazine on which he had consulted, declared that the *National Geographic* is "a magazine I've loved since childhood."[29] Even less adulatory references to the *National Geographic Magazine* indicate its wide circulation among modernist writers. As Stephanie Hawkins has discussed, Hemingway was not alone in his 1931 story "Homage to Switzerland," which poked fun at the magazine's efforts to romanticize masculine, heroic forms of travel and exploration and to treat those experiences as the basis for bonding among disparate cosmopolitans (190). These formulas became increasingly exaggerated and played out to the point that, by the 1930s, the magazine was the frequent target of parodies, perhaps none more biting than the *New Yorker*'s lampooning cartoons, as Hawkins's *American Iconographic* traces (see figure 1). Public recognition of these artificial qualities of the magazine's narratives about exploration and

"Geographic readers don't want that sort of thing, I tell you!"

Figure 1. Charles Addams cartoon from the June 29, 1935, issue of the *New Yorker*. © Charles Addams; printed with permission of the Tee and Charles Addams Foundation.

a celebrated notion of western cosmopolitanism can occur, however, alongside the geographic epistemological work that the magazine continued to perform.

A sense of the continued value of the magazine can be gleaned from the role that it played through the 1930s in the lending library of the well-known Parisian bookstore, Shakespeare and Company, which provided European and expatriate American modernists easy access to many of the *National Geographic*'s issues. Run by Sylvia Beach and Adrienne Monnier, Shakespeare and Company sold the latest in experimental fiction and poetry, hosted literary readings, and offered writers editorial help and an outlet for publication. (The bookstore is still well known today in part because it was the first to publish James Joyce's *Ulysses.*) Its lending library contained books and periodicals written in English that Beach selected and that circulated among its patrons (Beach 21). Borrowing books was a less expensive proposition for many in Beach and Monnier's orbit.[30] But Beach also put magazines into circulation, including multiple issues of the *National Geographic Magazine* spanning at least the 1930s and the early 1940s.[31] She also managed individual *National Geographic* subscriptions on behalf of the bookstore's patrons. André Gide, for instance, took out an individual subscription in the 1930s through Beach in order to support the bookstore during the difficult economic period following the great European and American financial collapse (Fitch 351).

Part of what the magazine offered modernist writers was a vital source of geopolitical information, despite the fact that the magazine's own editorial vision statement wanted the magazine to steer clear of anything potentially controversial. It also gave these writers material that they were able to make their own. A 1935 *National Geographic* article by Leo Roberts, "Traveling in the Highlands of Ethiopia," informed H.D. and Bryher of the nature of Italy's fascist, imperialist ambitions in the mid-1930s through its discussion of Mussolini's mounting interest in Ethiopia and Emperor Haile Selassie's looming ouster. It also highlighted elements of transnational and cross-cultural syncretism in the figure of Selassie himself that H.D. made a part of her psychoanalytic dream work of that period. Summing up in her correspondence the details of her dream imagery about Selassie, H.D. employs the shorthand that she and Bryher regularly used to refer to the things most intimate and familiar, and she refers to the *National Geographic Magazine* in this most offhanded of ways: "All this in Geographic and most interesting."[32]

In addition to geopolitical awareness, the *National Geographic Magazine* provided H.D. and other modernist writers with exposure to some of academic geography's signature ideas. In this section I examine some of the tendencies of the magazine throughout the first few decades of the twentieth century in order to trace the particular forms that environmental determinism took, codifying certain kinds of geographic epistemologies and styles for a middlebrow, mainstream audience. Although traditional narratives of the magazine tell the story of a schism between the world of expert academic geographers and an increasingly popular magazine for amateurs, this period saw considerable fluidity and exchange between academic and popularized forms of geography. Besides Semple, a number of other academic environmentalist geographers continued to be involved in the National Geographic Society despite the formation of a more "professional" academic society, the Association of American Geographers. For example, as Susan Schulten notes in *The Geographical Imagination in America, 1880–1950,* William Morris Davis founded the Association of American Geographers even while continuing his association with the National Geographic Society, giving lectures to the NGS and engaging in fund-raising on its behalf (50).[33]

Examining the forms of expression that environmental determinism took in the *National Geographic Magazine* helps to fill out aspects of the geographic epistemological work the magazine performed in its early decades. Schulten and others have discussed the ways that the environmental determinism in the magazine colluded with American investments in the Spanish-American War of 1898, although past this point is where most academic geographers and historians trail off, gesturing only vaguely to the pliability of environmental determinism in the post-1898 afterlife of the magazine.[34] Most of the scholarship on the periodical has not addressed its engagement with environmental determinism, instead focusing directly on the nationalist interests served by the interplay between the magazine's photo spreads and texts and the primitivist discourses in which such coverage of the non-West participated.[35] Closer attention to the articles published between 1900 and the 1930s reveals a persistent linkage between the natural environment and the nature of human society, one that exists within the textual form and layout of the magazine articles themselves. These articles signal the ways that environmental determinism came to serve not only as scientific content but also as a literary style of privileged geography, constructing it as something that precedes and enables human society rather than the other way around. If literary modernism benefited from the temporal orientations of Bergsonian ideas of inner time or Freudian narratives of identity formation, it also benefited from the primacy afforded to the physical environment.

The very earliest issues of *National Geographic Magazine* were rife with environmental determinist discourse at a time when the magazine was still largely a technical journal for the geological scientists who worked for the federal government and before the rise of academic geography in universities. Beginning in 1888, the year he formed the National Geographic Society, Gardiner G. Hubbard gave an annual lecture called "The Geographic Progress of Civilization." The seventh address, in 1894, maps the successes of human civilization, measured as they were by western standards, in terms of the benefits of natural environment that geographical location provided: "If parallels of latitude were drawn around the earth about fifteen degrees north and fifteen degrees south of Washington, the land within these parallels would include all the countries of the world that have been highly civilized and distinguished for art and science. . . . The nearer man lives to the polar regions the greater his inferiority in intellect, the greater his barbarism" (Hubbard,

"Geographic Progress" 1). Even a place viewed as fertile as the African continent produces only barbarism: "As the earth here spontaneously furnishes food for the sustenance of man, and as only scanty clothing is required, all inducements to either mental or manual labor are wanting" (16).

While the *National Geographic Magazine* advances a hardened version of Semple's manner of differentiating western, "advanced" civilizations from the "primitive" non-West based on environmental factors, it also echoes environmental determinism's comparativist logic that draws links across space and across cultural difference based on principles of environmental similarity. William Morris Davis's 1893 address to the National Geographic Society, published that same year in the *National Geographic Magazine* as "The Improvement of Geographical Teaching," encouraged educators to engage in side-by-side comparisons of comparable regions in various parts of the globe, with the idea that the same climate and topographical features will produce similar results (finding correspondence between plateaus in New England and along the German Rhine, for example [71]). The magazine's hopes for comparative geographic education fostered an "associational way of thinking" that crosses national borders, as Stephanie Hawkins has argued (36). I would root this manner of thinking in the underlying logic of environmental determinism, which assumed similar societal outcomes will come from similar natural environments.

One way that the magazine made environmentalist ideas usable to modernist writers was by codifying an environmentalist style that reflected in formal terms the assumption that the geographical environment is a causal, shaping force. As several geographers have explained, the legacies of environmental determinism shaped the very layout of geography textbooks. Formally, they reflect environmentalist causality, addressing geology and climate first, then moving sequentially to soil and vegetation, human settlement patterns and agriculture, and finally to industry and transportation systems.[36] I would extend this observation to claim that the *National Geographic Magazine* repeats this narrative grammar and transmits it, as well as its underlying logics, to a broader audience that included modernist writers.

This pattern typically informs *National Geographic*'s profiles of various global locales—that is, when the narrative isn't driven by the need to titillate readers with the exploits of a manly explorer/writer facing down multiple dangers, a figure well described by Hawkins. A summary of Harriet Chalmers Adams's 1922 essay on Salvador (today, El Salvador) offers a representative sampling of what readers of the magazine would routinely have encountered. The title alone suggests how features of the natural environment will be prioritized over economic, cultural, religious, or political arrangements: "Volcano-Girded Salvador: A Prosperous Central American State with the Densest Rural Population in the Western World." The essay begins with what appears to be an instant reorientation toward the human activities of Salvadorans, as it points out that Peruvian balsam as an export actually comes from Salvador (189). But the second paragraph returns the focus immediately to the geologic and botanical features of this "volcano-girded, forest-fringed country," and its accompanying photos draw the reader's eye to the topography of the environment, none more dramatically than a majestic photograph of Izalco, one of several Salvadoran volcanoes (see figure 2). Adams proceeds to provide an overall sense of its geographical location relative to nearby countries, its relation to oceanic waterways and its scale, and after a brief mention of the size of the country's population, she devotes extended attention to the country's two volcanoes and their relationship to earthquakes in the region. Only after discussing these geologic, topographic, and botanical features of El Salvador does Adams focus on human actions, first the country's agricultural and economic activities (its production of balsam and coffee) and finally matters of culture, religion, and politics, including the cultural and racial makeup of the Salvadoran population and its Spanish and Mayan origins. From there, the essay addresses music, food, economic infrastructure, and politics, adopting a patronizing attitude to this "fresh and smiling little country whose people form a limited aristocracy, cultured and hospitable" (200). The surprising "civilized" qualities to be found here, Adams suggests, stem from the many freedoms and privileges Salvadoran upper-class women share with the men of that country, which Adams describes in some detail. While it is tempting to claim that Adams adheres to an environmentalist narrative out of a need to bury her feminist observations about gender, other articles bear out this formula as well, including an essay in the same issue by Paul Popenoe, "Costa Rica, Land of the Banana."

Expatriate modernists would have encountered the same narrative structure on the other side of the Atlantic. Although England did not produce a direct equivalent of the *National Geographic Magazine,* in 1935 the Royal Geographical Society published the more academic and less widely read *Geographical Magazine,* and other middlebrow sources disseminated environmental determinist ideas about geography through their own blend of education and entertainment.[37] One of the clearest examples comes from the British photo-illustrated gazette *Countries of the World.* The well-known and

Figure 2. "Izalco, the Safety Valve of Salvador," from "Volcano-Girded Salvador" by Harriet Chalmers Adams, *National Geographic,* February 1922, 191. By permission of Harriet Chalmers Adams and National Geographic Creative.

prolific British editor J. A. Hammerton, who during World War I published the elaborate photo-rich weekly *War Illustrated,* produced just after the war a similarly lavish popular gazette, *Peoples of All Nations,* which conveyed anthropological ideas about various world populations for general readers. On the heels of its success, Hammerton published *Countries of the World* in order to bring geography to life for nonexperts. This was also a lavishly illustrated multi-part serial that offered essays arranged alphabetically on cities and countries all over the world. The periodical appeared in 1924 and 1925 (Stevens and Stape 203) and was subsequently reissued in book form.[38] Its contributors included academic geographers, travel writers, and literary modernists, which further unsettled the boundaries between geography as a science and literature and culture. Tellingly, Hammerton contracted Joseph Conrad to write an essay, "The Romance of Travel," as the lead article in the very first issue. This essay is now more familiar to American readers and cultural geographers as "Geography and Some Explorers," the title under which Conrad republished it first as a private pamphlet for collectors and then in America in *National Geographic* and several newspapers (Stevens and Stape 205).

Hammerton wanted to accomplish something similar to the *National Geographic's* efforts to captivate and instruct: to popularize geography by making it "humanly interesting" (Hammerton i), a middlebrow approach that meant balancing edification and entertainment.[39] By expressly appealing to a "wide and non-specialised audience," Hammerton could realize his vision that, no matter one's profession, "a knowledge of geography is essential to every man or woman with the slightest pretension to culture" (vi). One wonders if his inspiration was, in fact, the highly successful *National Geographic Magazine*, which took as its mandate revealing "the world and all that is in it" to the armchair reader at home and which had by the 1920s long attracted its wider readership in part through lavish photography. In his editorial statement, he explains that youths educated under the current geographical schema would benefit from "newer methods in making geography a living thing and something more than the memorizing of meaningless names." He goes on to assert, "One ought to know where Constantinople is, but equally important to know are the reasons why that particular part of the earth's surface has been for nearly two thousand years the inevitable site of one of the most important cities" (ii). Hammerton elaborates on this by agreeing with the suggestion offered by Sir Thomas Holdich, an "illustrious veteran of geography" who from 1916 to 1918 was president of the Royal Geographic Society (Baigent), that this "new geographical work, COUNTRIES OF THE WORLD, should [like Hammerton's previous gazette, *Peoples of All Nations*] also mark an epoch and deal not only with the geography of communications, but with the relation between the lands and their people" (ii). The geographers and travel writers supplying entries for the periodical thus have attended, he points out, first to the general environmental aspects of the country, next to the flora and fauna, followed by geology, then communication, transportation, and trade. Finally, the entries conclude by focusing on "the characteristics of the inhabitants, chiefly from the point of view of geographical influence upon them." Although individual writers were free to follow their own tastes, the formula makes it possible for them to "preserve all the merits of a scientific geographical work, while adding the attractive qualities of 'human interest' and variety of treatment" (iv).

Both *Countries of the World* and the *National Geographic Magazine* contribute to a narrative logic that enables key moves that modernist poets employed and that readers have been conditioned to understand: when landscapes are "presenced" on the page in literary texts, they operate as entry points to the cultural particularities of given regions or locales, through an assumed slide from environmental landscape to the social bodies that it enables and supports. Modernist poets manipulated this environmentalist formula of landscape first, culture second, so that when they annihilate space and bring conventionally distant geographical locales into contact on the page, the cultures they support "travel" with them. What this enables are new forms of transnational and transcultural affiliation and attachment that follow from and are made possible by new and specifically geographical arrangements. In *Trilogy*, H.D.'s expatriate speaker is able to access Egyptian myth and spiritual rejuvenation precisely because the physical geographies of London and Egypt's Karnak can blend into one physical environment.

As environmental determinism made its way into the 1930s, it became an increasingly contested and malleable set of ideas and took increasingly aestheticized and abstracted forms. Rather than dismiss environmental determinism outright as a theory that could no longer offer any explanatory purchase on the world, as academic geography had begun to do in the mid-1920s, *National Geographic* writers often invoked it to suggest that the United States had developed beyond or out of the reaches of environmental controls on society, while implying or stating outright that the rest of the world, less developed than the United States, continued to be subject to the controlling impact of the environment. This positioned American culture as superior in its creativity, force of will, and the unified drive of its democratic masses so as to win out over any limits or determining influences of the environment. In other words, the magazine subordinates environmental determinist rhetoric to American exceptionalism, and it provides American culture with a seemingly less overtly political means through which to construct its cultural and national superiority on the global stage. In broader terms, it signals the adaptability and mobility of environmentalist ideas that can be deployed in multiple ways as a style. But these deployments signal the increasing geopolitical stakes for how and under what circumstances environmental influence gets deployed: whether one is or is not subject to environmental influence becomes increasingly aligned with positions of civilizational inferiority or superiority.

Through its deployment of an environmentalist style, the implicit assumption that the *National Geographic Magazine* as a middlebrow magazine passes off educational knowledge as "objective" is that the United States is uniquely equipped to dominate the landscape and that the rest of the world is not equipped to do so. For example, Frederick Simpich's "This Giant That Is New York," published in 1930, trumpets the ingenuity and innovation that have produced contemporary New York City. Here, the city is characterized as "Man's incomparable feat! As incredible, almost, as that ants should have built the Andes!" (Simpich 517). In likening New York to famous monuments of the past, Simpich portrays past human achievements as scars on the planet's surface, subject to environmental decay. But not so New York City: "As each edifice melts into new lights and shadows, all the architectural phantasmagoria of the ages, linked even with earth's ancient scars, seems to unfold. In fancy you see a phantom city, a kaleidoscopic riot of Alps and Acropolis, Ming Tombs, Taj Mahal, and Pyramids; Gibraltar, Panama Canal, Tower of Babel" (517). The essay as a whole trades on an extended conceit in which human innovation and ingenuity have surpassed and taken over geological changes to the earth's face, and the moons, comets, and rainbows of nature have been replaced by the Manhattan skyline at night. Throughout, Simpich uses geological metaphors to describe the city, suggesting that the cause or responsibility for shaping the world as we know it has shifted from the earth to human hands. One photo of the New York City skyline of skyscrapers, taken from the perspective of Central Park's lake, is presented along with the caption "Man-made crags rise above Central Park Lake" (xx) Elsewhere, Simpich implies that human achievement has bested even the most spectacular example of American geomorphology when he calls Fifty-Second Street "The Grandest Canyon" (xxi).

Another essay from the 1930s, "New York—An Empire within a Republic" by William Joseph Showalter, one of *National Geographic's* frequent contributors, celebrates the wonders of New York productivity and innovation through a kind of flipped environmentalist language, in which agency is given over to the human side of the equation but in a way that uses the form of environmentalism and is couched in its language. Manhattan constitutes a "brilliant record of human achievement," but even in an essay that bombastically proclaims New York an empire on the same level as the Greek, Roman, and British Empires, he presents New York's achievements as direct results of the favorable influence of the environment: "New York's people have ever been ready to capitalize every advantage of geography. They built their chief city at the crossroads between New England and the seaboard Colonies farther south," which allowed the development of foreign shipping. Although the main emphasis is on human achievement, Showalter repeatedly uses geologic and natural language to mark these successes that conflate the environmental with the man-made: Manhattan's city skyline is described as having an "overpowering roar of the mighty man-made canyons of Manhattan" (Showalter 515). The people who work there arc like the "beavers that once roamed where they live" in the way they "serve varied needs of the Nation" (515). But rather than completely eclipsing environmentalist with human agency, the metaphors themselves work against the notion of a total reversal. Showalter likens human-produced innovation to features of the natural environment, suggesting that it is the bellwether and has not been completely displaced.

While environmental determinist discourse, in its initial academic form, largely minimizes human agency over the land, it gradually becomes the terms for a developmental narrative about American superiority, in which America could differentiate itself from other parts of the world "still stuck" in a mode in which the environment was the more dominant player. This particular deployment of an environmentalist style supports notions of American exceptionalism that modernist writers such as Langston Hughes and Gertrude Stein contended with. But as the following chapters will demonstrate, these and other aspects of environmental determinism's formulations are taken up by modernist writers increasingly far from geography's specific focus on the natural physical environment. Environmental determinist ideas take on lives of their own and become an increasingly adaptable and strategic discourse and aesthetic, even after academic geography moves on in the mid-1920s to embrace other models of the human-land relationship. Environmental determinism thus instantiates a method/structure of global comparison that helps to shape the terrain of the global projects of outward-looking American writers. This aspect of twentieth-century geography is not entirely different from the global aspects of connection we see embraced by nineteenth-century cosmological geography, just as the global American literary projects of Hughes, Stein, and H.D. are not all that different from Whitman's.

2

The "Terraqueous" Globe

Walt Whitman and the Cosmological Geography of Humboldt and Somerville

The roots of geopoetics, in which poetic formal experimentation enacts global connection and comparativism, can be found in Walt Whitman. His "Passage to India," composed and revised in the last decades of his life, provides an illustrative starting point for tracing the intersections between poetic experimentation by American writers and the geographic epistemologies that held sway for them. The poem is global in scope. As recent readings have discussed, it considers space-collapsing innovations of the late 1860s, in particular the construction of the Suez Canal abroad and the Pacific Railroad at home as well as Christopher Columbus's voyage to find the New World in search of a "passage to India," and it marshals a wide range of histories and geographies that are not contained within America's conventionally drawn spatial borders. But one significantly overlooked dimension of the poem is the debt it draws from the European cosmological geography of Alexander von Humboldt and Mary Somerville in its search for a compendious unity, which provides Whitman not only with a blueprint for truer forms of transnational connection that the poem seeks but also with disciplinary authority that allows him to "go global," ultimately in ways that exceed the power of the geographer himself or herself. Whitman thus helps to illustrate the long genealogy of the intertwined development of geographic knowledge and experimental modernism at the same time that his work highlights the ways that such braidings and unbraidings of the geographical and the literary are often accompanied for American writers by contests over literary authority. Moreover, Whitman declares a superlative space for the distinctly disjunctive, paratactic nature of experimental poetics to create forms of global connectivity in ways that later modernists develop in a variety of directions. Exactly what geography "authorizes" experimental poets to do is part of what this and succeeding chapters will trace.

Viewing "Passage to India" alongside Humboldt and Somerville illuminates the complex ways that American culture and geography intersected in the nineteenth century and raises the stakes for recognizing that connection, given that the politics of nation in nineteenth-century America was founded to a large degree on geography. As several American Studies scholars have observed, an appeal to certain American topographical features and geographic locales was a key technique for carving out a sense of national identity distinct from the established certitude of British and European cultures, and it is in this context that a wide array of nineteenth-century writers borrowed from geography texts, maps and map-reading techniques, travelogues, pamphlets, and land surveyance.[1] The reasons for America's interest in geography seem clear given the relative instability of America's own geographical makeup. In the period before the Civil War, the nation was plagued by an anxiety about geographic expansion, perceiving the incorporation of new territories as both "formless and threatening," as Anne Baker has characterized it (7). Not only were national divisions regarding slavery making it difficult for the nation to achieve a cohesive definition of itself, but the nation's borders on multiple fronts were not yet codified and mapped according to conventional standards: the ever-expanding western borders under the banner of manifest destiny, conflicts with Canada over the northern border, the annexation of Texas, and the new acquisition of Florida from Spain all created uncertainty about the literal form of the nation (Anne Baker 2). The production of atlases during the first half of the century reflected these national concerns, as evidenced by the popularity of texts such as Sidney Morse's 1849 *System of Geography for the Use of Schools: Illustrated with More than Fifty Cerographic Maps and Numerous Wood-Cut Engravings*. By the end of the nineteenth century, however, national interests shifted from a focus on internal domestic geography to that of world comparison (Schulten 17). During the late 1860s in the lead up to the publication of "Passage to India," Whitman seems to anticipate this shift from an intense fascination with national geographic consolidation to a more outward looking internationalist way of conceiving of geography, and he routes this orientation through European cosmological geography.

It is only recently that scholarly attention in an Anglo-American context has recognized the international dimensions of Whitman's work.[2] He is in many ways one of the most firmly and self-consciously American-bound poets, whose nationalism certainly rings true in the groundbreaking early editions of *Leaves of Grass* of the mid-1850s and early 1860s. Later poems such as "Passage to India," however, tell a different story. In part because of recent developments in postcolonial and American Studies methods, the critical conversation about Whitman's work has been able to move past dismissals that in the 1970s viewed the later work as having taken a disappointingly wrong turn into the realm of philosophy, religion, and abstraction.[3] "Passage to India" has now come to serve as a significant flashpoint for discussion about whether Whitman supports or critiques American exceptionalism and proto-imperialism.[4] It is necessary to include Whitman's relationship to cosmological geography in such debate; the conception of a unified globe as well as cosmological geography's orientation away from man-made political divisions and social forms of belonging have direct consequences for our understanding of the poem's national and/or transnational politics.

In one of his many conversations with his disciple and amanuensis Horace Traubel, Whitman explained his thinking about "Passage to India." "There's more of me, the essential ultimate me, in that than in any of the poems. There is no philosophy, consistent or inconsistent, in that poem . . . but the burden of it is evolution—the one thing escaping the other—the unfolding of cosmic purposes" (Traubel 1: 156–57). If the poem ostensibly contemplates the completion in 1869 of the Suez Canal and the Pacific Railroad, which then prompts a backward look at the space-traversing technologies of Christopher Columbus, who sought a "passage to India" but "discovered" the Americas instead, beneath these layers the poem presents an "evolution" that reveals the "cosmic purposes" of the geography of the universe. Whitman's belief that the underlying primal movement or plan of the universe reveals itself through the evolution of physical matter, perceivable through the active pursuit of understanding that process, shares an uncanny resemblance to the work of European nineteenth-century cosmological geography. Most visibly identified with Humboldt and Somerville, this dominant paradigm of physical geography produced beyond the Atlantic approached the entire globe as an interconnected system of individual parts working in harmonious collaboration and which expressed some larger design at work. Indeed, in the notebook in which Whitman devised the "ground plan or the sky plan" of the poem, to cite Traubel's transcription, Whitman explicitly mentions the process of mapping geographical features; he hoped to incorporate painting a picture of Columbus's voyage of discovery and then "[i]n the course of the piece, a geographical and other description of the country through which the Continental Railroad passes in the States, (then names) the fauna, the mountains, rivers, &c." (Traubel 4: 399). Given Whitman's habit of researching the subject matter of his poems, which in the past had included consulting geographical materials,[5] a poem with the comprehensive geographical scope of "Passage to India" and that proceeded *as* a geographical description would naturally have led Whitman to Humboldt and Somerville, whose work by the late 1860s had achieved singular prominence for dealing with geography in "global" proportions.

Although certain aspects of Whitman's connection to Humboldt have been discussed on occasion, it is surprising that a systematic study of Whitman's relationship to Humboldt or to this brand of European cosmological geography in general has not yet been undertaken.[6] The focus has mainly centered on the obvious overlap between Whitman's "Kosmos," one of the many poems added to the 1860 *Leaves of Grass*, and Humboldt's own signature work, *Cosmos* (with Whitman even alternating between the English and Humboldt's German spellings of the word in various versions of the poem).[7] In the poem's *in medias res* beginning, Whitman personifies the cosmos as a democratically absorptive entity and celebrates its ability to encompass everything, an inclusiveness highlighted by his use of the "and": "Who includes diversity and is Nature, / Who is the amplitude of the earth, and the coarseness and sexuality of the earth, and the great charity of the earth, and the equilibrium also" (1-2).[8] The imagery directly invokes Humboldt's belief that the universe is one harmonious whole, rather than assuming that we live in a universe always on the verge of anarchy or in a constant state of being pulled apart by the vast diversity of life forms (Reynolds 244). Indeed, the poem celebrates the power of Humboldt's geographical theory to comprehend these aspects of the natural world: "Who, out of the theory of the earth and of his or her body understands by subtle analogies all other theories, / The theory of a city, a poem, and of the large politics of these States" (7-8). In this set of images we can identify geography, the theory of the earth, as the scientific substrate upon which "all other theories" emerge; we can also identify the ways that Whitman's disjunctive poetics—here set off by the anaphoric catalog of lines that start with "who"—blur the boundaries between the cosmological geographer and Whitman the poet who understands not only "all other theories" but the "theory of a city, a poem, and of the large politics of these States."

Although Humboldt's influence in nineteenth-century America is undeniable, Whitman's engagement with academic geography produced on the other side of the Atlantic does not end with this figure alone. Previously overlooked evidence indicates Whitman's familiarity with Mary Somerville's 1848 *Physical Geography*, which echoes Humboldt's belief in a harmonious, ordered universe but expresses those ideas in lyrical, romantic, and explicitly religious terms, qualities of her writing which the prominent periodicals reviewing her work foregrounded and even exaggerated. Whitman clipped out a lengthy review of Somerville's *Physical Geography* published in the *North British Review* the same year the book came out.[9] Although unsigned, the review was written by Sir David Brewster, who was, like Somerville, from Jedburgh, Scotland, and whose background as a physicist, astronomer, mathematician, and teacher allowed him to engage deeply with her work.[10] The early date of this magazine clipping might seem to limit the temporal boundaries of Somerville's influence, but American periodical coverage of Somerville's achievements, not to mention Whitman's own manner of revisiting his magazine clippings at later dates, would have easily kept her in view.[11] In addition to frequent reviews appearing in popular magazines like *Godey's Lady's Book*, feature articles appeared in periodicals like *Atlantic Monthly*, which in 1860 published a glowing essay on Somerville that grouped her with Elizabeth Barrett Browning and painter Rosa Bonheur as one of the "few women of genius" who have successfully rivaled their male counterparts ("Mary Somerville"). In 1869, Somerville was awarded not only the prestigious Patron's Victoria Medal of the Royal Geographic Society but also the first gold medal of the Geographical Society of Florence (Sanderson 418).

It is no surprise, then, that one of the most distinctive lexical choices Whitman makes in "Passage to India" refers to work by and about Somerville. In the middle of the poem, the poet has a vision of a "vast terraqueous globe" (120). This highly unusual adjective appears nowhere else in Whitman's published work, but the phrase "terraqueous mass" (76) does appear in the Brewster review of Somerville's work that Whitman had preserved. Brewster uses the phrase to characterize the liquid and primordial world geography Somerville conveys, and that kind of undifferentiated amorphous spatial expanse is exactly what "Passage to India" seeks and tries to enact.

One of the reasons that the currency of Humboldt and Somerville's cosmological geography should not be seen as limited to the early part of Whitman's career is that he considered science to be the bedrock support for the poet and a wellspring from which the poet could continually draw. In his preface to the 1855 *Leaves of Grass*, Whitman proclaims:

> Exact science and its practical movements are no checks on the greatest poet but always his encouragement and support. The outset and remembrance are there . . . there the arms that lifted him first and brace him best . . . there he returns after all his goings and comings. (626)

The preface goes on to list a number of other professions that provide fodder for the poet, but Whitman singles out science as primary and superlative. Although to modern-day readers this category of "science" might seem to encompass fairly narrowly the "hard sciences," Whitman's invocation would certainly have been understood by his contemporary readers to include geography. As I discuss in chapter 1, not only was academic geography classified in the nineteenth century as science of the first order, particularly in European contexts, but because of its synthetic incorporation of geology, hydrology, botany, physics, astronomy, and the like, it was also perceived as the "mother of all disciplines." If science operates in Whitman's preface as a maternal embrace to which the poet can repeatedly come home, geography certainly promised many returns to the poet over a long career.

My argument here is that "Passage to India" mines European science, and cosmological geography in particular, in multiple ways and uses it as the ultimate route to organizing the globally networked system the poet seeks and tries to produce. If, as Hsuan Hsu has argued, the poem "jumps scales" in leaping from the local to the national to the planetary as a response to global capitalist expansionism, it is cosmological geography that provides a multi-scalar model for conceiving of these interconnections in terms that are distinctly *not* organized around national-imperio-economic axes. In fact, it is through the disciplinary perspective of academic geography itself that Whitman authorizes his own poetic project and then exceeds it, becoming in "Passage to India" the ur geographer-poet who can reveal the divine at work within the planetary through his pursuit of understanding the natural world.

This perspective allows us to see "Passage to India" actively embracing global unity as an index to the deeper cosmic purposes of the universe, which cosmological geography tenders as a realizable promise, instead of assuming the poem's pessimism and its resignation about the failings of human attempts to create a truly democratic global highway, as imperio-capitalist readings have done.[12] The "Passage to India" that Whitman constructs in his poem can be read, then, not as a successful or failed actual route to India, nor as a passage to the United States as a proxy *for* India, but instead as a commitment to pursuing the recesses of a "cosmic" plan that cosmological geography can unlock and that provide, in turn, a diffuse, global sense of cosmic citizenship.

In this latter sense, recognizing the poem's investments in cosmological geography as the delivery system for non-national belonging necessarily alters our sense of the geopolitical compass of the poem. In working out of this nineteenth-century vein of geography, Whitman embraced the work of scientists who moved away from nation-state thinking and its extended reach through imperialist expansion and capitalist trade, a set of principles with which American readers would have been familiar given the stature of Humboldt and Somerville on the domestic front. Although "Passage to India" in some measure bends the extra-national cosmopolitan sensibility of cosmological geography to serve American interests, in other respects this sort of orientation serves as a counterpoint to the rising tide of American exceptionalist ideas in the nineteenth century and the promise of imperialist and neoimperialist forms of global trade that would enrich the nation.

Whitman considered Humboldt to be one of the most important figures of his time. He kept magazine clippings related to Humboldt,[13] and rumor has it that he wrote *Leaves of Grass* with a copy of Humboldt's *Cosmos* on his desk.[14] He was also keenly aware of Humboldt's death in 1859, recalling the date much later with perfect clarity, as Traubel relays: "The mention of Humboldt caused the remark that Humboldt, Macaulay and [Washington] Irving died the same year (1859)—'And Humboldt head and shoulders above both the others' W. thinks—'easily—easily. And the brother, too—Wilhelm—a great man by all my means of knowing. Sometimes they hunted *double*—in pairs, I think'" (Traubel 6: 309). Although the comparison of Humboldt to Macaulay and Irving is initially triggered by accident of their dates of death, in measuring Humboldt against two figures at the height of their fields—Thomas Babington Macaulay in history, Washington Irving in literature—Whitman finds Humboldt far superior to both.

Humboldt's death in 1859 in fact may well have informed Whitman's decision to incorporate a globe into the visual design of his 1860 edition of *Leaves of Grass*, published less than a year after Humboldt's passing. This edition marks a stark visual departure from the botanical images incorporated into the cover designs of the earlier editions of *Leaves of Grass*. Instead of the leaves and grass embossed on the cover of the 1855 *Leaves of Grass* and sprouting in gold from the letters of its title, or the embossed foliage that adorns the green cover of the 1856 edition, the 1860 cover displays an embossed image of a globe featuring the Americas in the clouds, floating on the cover between the "Leaves" and "Grass" of the title (figure 3). This edition also features on its spine an image in gold of a butterfly on a hand as well as an embossed image on the back cover of a sun setting on a sea (figure 4). All three of these images also appear at various intervals within the pages of *Leaves of Grass*; the first instance of these is the butterfly image, which appears at the end of the table of contents (figure 5). Several scholars have discussed the globe and sunset as images of national uncertainty related to the brink of Civil War, although no theories have been offered about how Whitman happened to have chosen the images themselves, nor have scholars connected these images to the project of cosmological geography.[15] And yet these images align closely with the work of Humboldt in several key ways: the butterfly suggests his work as a naturalist, the featuring of the Americas on the globe evokes his extended scientific expedition in Latin America at the turn of the nineteenth century (indeed, he was thought of as the second "Columbus," as Buttimer discusses [viii]), and the globe itself as well as the sunset resonate with the principles of climatology, the tempo of the calendar, and the astronomical relations of the earth to other heavenly bodies that Humboldt's *Cosmos* detailed extensively. Given the proximity of Humboldt's death to the printing of the 1860 edition of *Leaves of Grass*, Whitman could very well have used these images to compensate for two forms of lack—the loss of certainty and unity in the future of the nation, and the loss to the world of ideas of a great cosmographer.

Whitman did more than admire Humboldt as an influential figure. As Traubel records, he identified with Humboldt directly and attempted to embody him in the quotidian rhythms of his life:

> W. suddenly took a notion to get up. I helped him to the chair. His legs are little good. He leans heavily on you. Yet he on his own plane is comfortable just now. "One thing is gone utterly and forever—my agility," he said as we walked across the room. Sat down. Stirred the fire. "I will get you to hand me the poker," he said. Then worked for fully ten minutes—likes it—with the embers, talking meanwhile leisurely and at perfect ease. Turned up light, too: brushed his hair back from his face and brow. Did he nap it always so [*sic*] in the evening as Ed said? "No: I have no rule: I live, move, just as the spirit directs." He had read somewhere of Humboldt's informal mode of life while in Paris—eating, sleeping, &c., not by hours but by instinct. W. liked the idea. Spoke of it. (Traubel 3: 446)

Figure 3. Front cover of the 1860 edition of *Leaves of Grass*. From the Trent Collection of Whitmaniana (Trent I-2, no. 4), David M. Rubenstein Rare Book and Manuscript Library, Duke University.

Figure 4. Back cover of the 1860 edition of *Leaves of Grass*. From the Trent Collection of Whitmaniana (Trent I-2, no. 4), David M. Rubenstein Rare Book and Manuscript Library, Duke University.

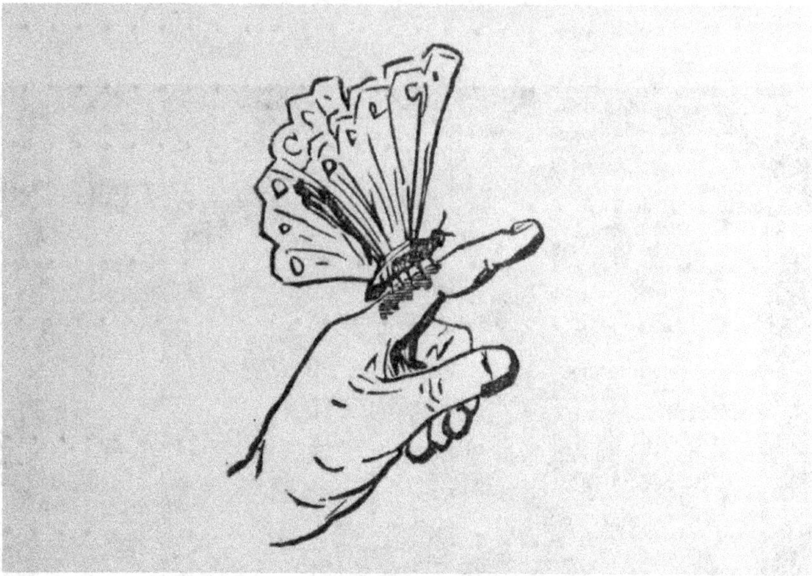

Figure 5. End of the table of contents in the 1860 edition of *Leaves of Grass*. From the Trent Collection of Whitmaniana (Trent I-2, no. 4), David M. Rubenstein Rare Book and Manuscript Library, Duke University.

Not only does this anecdote suggest Whitman's extensive familiarity with Humboldt, but it also crafts a kinship or even a shared subjectivity that allows Whitman to enact Humboldian ideas about nature and the cosmos in poems like "Passage to India" and move beyond them.

These dynamics of identification function similarly to the ways that Lincoln served as a political double for Whitman or even as an artistic alter ego that Whitman claimed as his own, in "When Lilacs Last in the Dooryard Bloomed."[16] Just as Whitman never actually met Lincoln in person as far as we know, he was never personally acquainted with Humboldt, and the composition of "Passage to India" occurs several years after the geographer's death. But Whitman was nevertheless highly capable of investing in poetically generative forms of homosociality across social distance and despite or perhaps even because of death (Pollak 164). Such dynamics unfold in the poetry not through overt reference but through subtle activation of his readers' commonly understood knowledge and context; that famous elegy to Lincoln does not contain any overt evocations of Lincoln, as the title of its cluster in *Leaves of Grass*, "Memories of President Lincoln," might otherwise indicate, precisely because it does not need to. Instead, it contains veiled and metaphorical references to Lincoln while enacting his vision and presence. Humboldt, the "Einstein" of the nineteenth century, may well be the evoked but largely absent presence in "Passage to India" for similar reasons. One review of Whitman's 1860 *Leaves of Grass*, for example, uncannily enough, compares Whitman to Humboldt in an offhand way that suggests how eminently familiar he was to the American public:

> Mr. Whitman's book may not be understood at all for a long time. Nature and Art alike often seem chaotic and incomprehensible, must receive patient consideration before they render up their meanings, and wait for ages for their interpretors [*sic*]. The physical cosmos cooled its heels a good while before HUMBOLDT arrived. It took two hundred years to understand SHAKESPEARE. (O'Connor)

In 1869, the year that anchors "Passage to India," Humboldt would have been on the mind of Whitman and his readers especially. For the centenary of Humboldt's birth, cities across the United States held festivities of all kinds ranging from parades to concerts to public eulogies (Buttimer viii).

This evidence makes a compelling case for the ongoing significance of Humboldt in much of Whitman's later writing. In "Passage to India" in particular, the allusions to Humboldt are subtle and suggestive rather than overt both because of the strength of Humboldt's immediate cultural presence for Whitman and his readers and because the entire poem operates as its own sort of Humboldtian project. Humboldt's name only appears once in "Passage to India" in the form of a reference to the eponymous "Humboldt range" of mountains in Nevada (58), which were named after the geographer by topographical engineer John C. Frémont. Though small, the reference enacts on the page a figurative stamping of Humboldt on the natural environment, on Whitman's own national soil no less. Other references function more obliquely. The poet in "Passage to India" envisions the end of his journey as a union between brothers, an image Whitman uses elsewhere to express the union between the poet's soul and God. Here we can also trace Whitman's palpable identification with Humboldt: the image simultaneously embodies a union of the poet's soul with Humboldt as a predeceased elder brother of sorts:[17] "As fill'd with friendship, love complete, the Elder Brother found, / The Younger melts in fondness in his arms" (222–23). This image of brotherhood also doubles for a different configuration of brothers: Humboldt and his own elder brother, the well-known linguist Wilhelm von Humboldt, whose work Whitman also knew. Wilhelm had died in 1835. That Whitman remarks upon their brotherhood in the comments to Traubel about Humboldt discussed earlier (Whitman noted that they hunted together, "double" or "in pairs") makes this connection an intriguing possibility.

Most of all, though, Whitman's pursuit in the poem for what he variously calls "O vast Rondure" (81) or the "vast terraqueous globe" (120) not only works from Humboldt's understanding of cosmological geography but also moves beyond it. From this perspective, "Passage to India" measures forms of global unity produced by scientific innovation and exploration against cosmological geography and finds them to be lacking. It is cosmological geography itself—both in terms of its disciplinary and methodological powers as well as its capability to ideologically and aesthetically resist nationalism and imperialism—that ultimately fuels the poet's ability to generate his own global vision. Whitman's turn to the natural world in the poem is not a retreat into the mystical in the face of failed forms of human unity, as Hsu suggests, but rather a calculated and purposeful one in which the poet merges with, and ultimately supersedes, the authority and power of the geographer.

The poem's opening is launched by a celebration of the first of these forms, the space-traversing innovations of the Suez Canal and the Pacific Railroad, which Whitman seems at various turns to wholeheartedly endorse. Both of these public works projects, the "modern wonders" of engineering, were completed in 1869, with the poem also implicitly referring to the laying of the Atlantic cable in 1866. In this first section, these developments are seen as having "outvied" (4) the seven wonders of the world: "Singing my days / Singing the great achievements of the present, / Singing the strong light works of engineers, / Our modern wonders" (1–4). This prompts a vision the poet has of "tableaus twain" in section 3 that collapse or conquer spatial expanse:

> Lo soul for thee of tableaus twain,
> I see in one the Suez canal initiated, open'd,
> I see the procession of steamships, the Empress Eugenie's
>   leading the van,
> I mark from on deck the strange landscape, the pure sky,
>   the level sand in the distance,
> I pass swiftly the picturesque groups, the workmen
>   gather'd,
> The gigantic dredging machines. (42–47)

This vision is immediately followed by one of the Pacific Railroad, which he sees "surmounting every barrier" (49). The train not only links disparate American topography ranging from the Laramie plains to the Humboldt mountain range to the western desert but it instantiates a global network that links Europe to Asia: "Bridging the three or four thousand miles of land travel, / Tying the Eastern to the Western sea, / The road between Europe and Asia" (62–64). Hsuan Hsu has discussed these as manifestations of the sort of large-scale "public works" ostensibly meant to contribute to a democratically accessible sort of good that would benefit all, but that in reality create uneven geopolitical and economic terrain with some benefiting more than others. Hsu cites American anxieties about the Suez Canal that promised to advance European access to the Pacific in ways that could outstrip what its own Pacific Railroad might provide in the way of an American Pacific empire (148), and certainly in the image of the Suez Canal we see French imperialist interests as a dominant force given that the steamship of Napoleon III's wife, the empress Eugenie, is "leading the van." For these reasons and others, Hsu argues that Whitman compensates for the limitations of such human projects of "global rondure" by substituting a romanticized, abstract "ideal of rondure for concrete struggles against and among imperial powers in the Pacific" (149). However, it is equally important to note that the homegrown variety of public works projects flatten American space as well as flatten the globe for others. The Pacific railroad surmounts "every barrier" on the domestic front but also performs the same function globally by creating a "road between Europe and Asia." In terms of the scope of spatial territory it links and its potential to allow the fullest form of democratic access, the Pacific Railroad can be seen as rivaling the Suez Canal, not just because of its surefire domestic benefits but also because of its simultaneous benefit for extraterritorial nations who will be brought into contact through transportation and trade.

Whitman also explores global connectivity in the form of Columbus's explorations of 1492, implicitly linking human exploration as one technology to the engineering of public works as another. Part of this is triggered by the common project of imperial economic expansion that drove both sets of endeavors; Columbus represents Europe's long-standing quest for access to India in the age of exploration in order to increase global trade and the economic and geopolitical dominance that would come with it, and this is a goal that the public works projects of 1869 also seek to realize in their own fashion. Whitman initially endows Columbus with the heroic qualities of "courage, action, faith" that seem to stand the test of time. The language of performance in the poem constructs Columbus as an actor worthy of conquering the drama of history itself:

As the chief histrion,
Down to the footlights walks in some great scena,
Dominating the rest I see the Admiral himself,
(History's type of courage, action, faith,)
Behold him sail from Palos leading his little fleet . . . (152–56)

The celebratory rhetoric, however, is not complete; Whitman goes on to highlight Columbus's failure, frustration, and tragedy: "His misfortunes, calumniators, behold him a prisoner, chain'd, / Behold his dejection, poverty, death" (158–59). Rather than offer up Columbus as a wholesale role model of global rondure in action or its opposite, Whitman frames him as one instance of manufactured globality on which modern civilization has seized. Columbus is, after all, "History's *type* of courage, action, faith," a parenthetical insertion that underscores Columbian exploration and imperialist expansion as more misguided historical construction than bellwether. History requires an agent like Whitman, who stands there "noting the efforts of heroes" (160), in order to take the better measure and the longer view of what history typically records as civilization's major achievements. The work of "heroes" is therefore mere deferral, which makes Whitman impatiently wonder in an extended parenthetical: "Lies the seed unreck'd for centuries in the ground? lo, to God's due occasion, / Uprising in the night, it sprouts, blooms, / And fills the earth with use and beauty" (162–64). The implication is that the poet, not Columbus or the way history writes the record of achievement, can seize the potential that still lies untapped and realize it in fuller form.

Unlike the potential of technological innovation or of exploration and discovery, the poet alone can realize the global unity the poem seeks. In section 5, Whitman frames the technological innovations of 1869 and from the legacies of Columbus's voyage to which those innovations are linked as that which cannot be justified in the sense of demonstrating their rightness or rendering them to his readers: "Then not your deeds only O voyagers, O scientists and inventors, shall be justified" (106). Their failure produces fretting hearts that need to be "sooth'd" (107) with a "secret" that "shall be told" (108), presumably by the poet, because it has not been revealed through these previous efforts. What remains is a world full of divisions and gaps: "All these separations and gaps shall be taken up and hook'd and link'd together, / The whole earth, this cold, impassive, voiceless earth, shall be completely justified" (109–10). What is key is that the necessary task of "justifying" the "whole earth" explicitly invokes the work of the cosmological geographer who "justifies" the earth in many senses of the word: proving or verifying its underlying principles in what would otherwise seem like random occurrences, making it legitimate as an object of knowledge, and demonstrating its divinity, free of sin (especially in light of the more religiously inflected cosmological geography of Somerville).

It is the poet, however, who can fully embody all of the senses of justification that even a powerful figure like Humboldt cannot. If we consider the meaning of "justified" in relation to printing, in the sense of justifying type, then it is truly only the poet who can "justify" the world in its global unity.[18] Here, Whitman draws upon his many years of experience setting type in his newspaper work and of active involvement in the typesetting and printing of his editions of *Leaves of Grass*. Thus the project of explaining, legitimizing, sanctifying, and *writing* the earth "shall be completely justified" by the poet rather than only partially so in only some of its senses. On a material and performative level, Whitman's own text "justifies" global unity by aligning the text on the page in ways that Humboldt cannot. It implicitly distinguishes between "writing" and the "deeds only" of the voyagers, scientists, and inventors. The divine power invested in the poet in this part of the poem has long been recognized by scholars, readings that are anchored in Whitman's bald declaration of being "the true son of God, the poet" (111). But the specific form of authority that Whitman seizes borrows heavily from the conventions and expectations of cosmological geography. After his declaration of poetic divinity, Whitman states the certainty of his ability to traverse the globe in ways that cosmological geographers like Humboldt did in the service of scientific "purpose":

> (He shall indeed pass the straits and conquer the
>   mountains,
> He shall double the cape of Good Hope to some purpose,)
> Nature and Man shall be disjoin'd and diffused no more,
> The true son of God shall absolutely fuse them. (112–15)

While this image of exploration may refer on some level back to Columbus, signaled by the reference earlier in the poem to "voyagers," it converges here with aspects of Humboldt's own South American expedition. The immediate concern with bringing together "Nature and Man" is central to the project of cosmological geography alone, as both Humboldt and Somerville posited a human element existing in harmony with nature. In practical and methodological terms, however, neither geographer tended to devote extended attention to the human element, preferring instead to describe the natural world and its underlying mechanisms and devoting attention to human life separate from the analysis of the natural world. Whitman not only generates forms of global unity in "Passage to India" that complete aspects of the journey to which he recommits at the end of the poem, but he outdoes cosmological geography by offering truer forms of union between the world and the human than cosmological geography can accomplish. It is only through the poet that "Nature and Man shall be disjoin'd and diffused no more" (115).

Although the beginning of section 6 seems to return to celebrating human-generated forms of globality, we can read these images as a direct product of the geographer's-eye view of the poet. Here, instead of mere connection we start to see images of merger, figured in the romantic language of a planetary marriage embrace. These images of blending and merging stem, significantly, from the poet's own ability to "see" them in this way:

> Year of the marriage of continents, climates and oceans!
>
> . . . .
>
> I see O year in you the vast terraqueous globe given and giving all,
> Europe to Asia, Africa join'd, and they to the New World,
> The lands, geographies, dancing before you, holding a festival garland,
> As brides and bridegrooms hand in hand. (118–23)

The last sections of "Passage to India" portray the poet embarking upon his own voyage, pursuing his own experience of the "terraqueous globe," rather than merely witnessing globalizing effects secondhand. In this respect the poem depends upon an unmediated access to the different sense of the cosmos that cosmological geography provides, one that reorients the poem away from the national and geopolitical issues linked to human technologies of exploration and engineering. It is, above all, a particular form of the terraqueous that the poet can take charge of as a Humboldtian figure himself.

> O soul, repressless, I with thee and thou with me,
> Thy circumnavigation of the world begin,
>
> . . . .
>
> Back, back to wisdom's birth, to innocent intuitions,
> Again with fair creation. (169–70, 173–74)

Whitman returns not only to the "wisdom's birth" or "innocent intuitions" of pure science but to the underlying purity and innocence of the cosmos itself.

Notably, this constitutes a different kind of terraqueous formation that does not suture artificially bounded and demarcated spaces together but rather moves through undifferentiated, non-national space. The poet sails upon seas that are "trackless" (177) and through regions that are "infinite" (190), and finally contemplates the sheer "vastness of Space" (211) in ways that contain time, death, and the entire universe. The poet explores the nested interconnections among aspects of the natural universe at all levels of scale that Humboldt and Somerville take on in their projects, encompassing hydrography, topography, physics, and astronomy, among other areas of scientific knowledge.

A significant part of the poet's journey involves accessing the divine principles that lay beneath the surface effects of the natural world, and for this, too, Whitman engages with cosmological geography, although not necessarily Humboldt's own brand of it. Somerville provides a model of cosmological geography that overtly links the natural world to the divine in ways that align with the final sections of "Passage to India" as well as with other of his spiritually focused, late-career works. Although Reynolds views Whitman as partaking in Humboldt's agnosticism (580), so much of "Passage to India" concerns the soul in ways that directly evoke divinity, however heterodox that might be in Whitman's hands, and in a manner that complements Somerville's own attention to the role of the divine. Land and seas, traversed by explorers, cosmological geographers, or inventors, are not enough to fulfill the needs of the soul and to reach "realms of budding bibles" (168). Although Somerville herself was censured by religious authorities for the way she presents religion in her work,[19] *Physical Geography*'s explicit religiosity contrasts with the much more muted strains we find in Humboldt's work (Neeley 229). Her focus on the operations of a divine principle partly stems from her own particular geographic and academic location; as David Livingstone has explained, scientific work in Scotland in the eighteenth century was invested in the interplay between natural science and moral philosophy and couched this linkage as an expression of some divinely orchestrated plan (120). Whitman's preliminary ideas for "Passage to India" incorporate religion quite purposefully, as an exploratory sketch for the poem in his Lion notebook indicates.[20] He planned all along to conclude the poem with a turn toward the larger cosmic design for the universe, as the cosmological geographer might, which involves a distinctly religious component that completes what is only potential in history's explorers: "The spinal idea / That the divine efforts of heroes, & their ideas, faithfully lived up to. Will finally prevail and be accomplished however long deferred." Whitman manifests the religious thematic of this plan in part by incorporating poem fragments from the Lion notebook into his composition process, the most notable of which is "O Soul, thou pleasest me" in which the poet is "Chanting my chant of God in / silent thought, / O Soul, thou pleasest me—I thee." If earlier in "Passage to India" technological innovation has made us "feverish" and "unsatisfied" of soul (91–92), more direct contemplation of the cosmos as "God's handiwork" can remedy that problem. Early on in the poem nature is "impassive" (95), "separate" and "unnatural" (96) and "unloving" (97). But by the time we reach the poem's conclusion, the human, the natural, and the divine have been brought into harmony, as Somerville's cosmological geography would have it.

What, then, does attention to Whitman's engagements with cosmological geography bring to the poem's political orientation? In her brief discussion of the Humboldtian connections in "Passage to India," Laura Dassow Walls contends that Whitman invokes the universe of cosmological geography but "places" it squarely on American soil, thereby bringing the entire globe into the figurative possession of America. Her claim hinges upon Whitman's declaration of "my land" in the poem's final section, which she reads as an expression of American exceptionalism and manifest destiny:

> Passage to more than India!
> O secret of the earth and sky!
> Of you O waters of the Sea! O winding creeks and
>   rivers!
> Of you O woods and fields! of you strong mountains of
>   my land!
> Of you O prairies! of you gray rocks!
> O morning red! O clouds! O rain and snows!
> O day and night, passage to you!
>
> O sun and moon and all you stars! Sirius and Jupiter!
> Passage to you! (233–41)

The topographical features of prairies, woods, fields, and the "strong mountains of my land" seem to stake a distinctly American space as an anchor for the poet, which then in Walls's reading would tie things partaking in the higher order unity of the cosmos, the "secret of the earth and sky," to America. The formal strategies at work here as well as the poem's final stanza complicates this reading, however; the poet looks beyond these geographical and political borders in a centrifugal gesture that embraces cosmic oneness through both form and content.

Whitman creates a kind of cosmological formal structure in this passage that overrides individual, fixed spatial emplacements or circumscriptions of national location. His anaphoric repetition of "Of you" and "O" is a strategy of juxtaposition which lays side by side a variety of geographies and climates, shifting jarringly from one kind of scale of orientation to another: from particular creeks, rivers, and mountains to something as big as the solar system. The "O" that invokes these various components of the known universe is also the "O" of cosmological Oneness that is created when Whitman's experimental form brings them into relation. The open nature of parataxis prevents us from ascribing special significance to distinctly American environments that appear in the stanza or from constructing a narrative of a cosmological journey that begins in or is structured by a decidedly American point of origin. Although the project of the cosmic divine is not yet realized by the poem, Whitman's cosmological geopoetics fuel his concluding exclamations of confidence and his assertions of safety: "O daring joy, but safe! are they not all the seas of God? / O farther, farther, farther sail!" (254–55). Only through the poet can the reader sail into these images of undefined, undifferentiated spaces of the divine with exuberance, unlike mediated forms like that of academic geography that only offer stupefaction: "Have we not darken'd and dazed ourselves with books long enough?" (247).

The second half of "Passage to India" may look like a turn to the metaphysical as a form of retreat, or like a simple shift toward considerations of the divine in the natural that animate much of Whitman's earlier poetry, as scholars have suggested. Yet there is a textual geopolitics to Whitman's trajectory, a distinct line of sight offered by cosmological geography that overtly privileges the natural as an expression of the divine and as a way to purposefully look beyond or beneath the violence associated with nationalist formations of the human. As chapter 1 discusses, Somerville's *Physical Geography* explicitly states that it will look past the artificial boundaries of the human. Brewster's *North British Review* of Somerville highlights the clearly divisive nature of human politics that not only erects arbitrary boundaries but does an injustice to nature's boundaries:

> From our youth we have been accustomed to look at the Earth, or its delineations, as mapped onto regions, from which the great boundaries of nature are effaced. Empires purchased by blood, and held by force, are, in the political geography with which we are familiar, bounded by chains of custom-houses and barriers of forts. Ambition has replaced the sea-line, and the river, and the mountain range, with frowning battlements, cordons of troops, and rapacious agents—parceling out the earth into unnatural divisions—forcing its population into jarring communities—severing the ties of language and religion—breaking up into hostile principalities the fatherlands of united hearts—extirpating even the native possessors of the soil, and thus treating intellectual and immortal man as if he were but the property and the tool of the tyrant. (Brewster 77)

To embrace the global fluidity of cosmological geography is to self-consciously reject the violence generated by humanity's manner of erecting artificial divisions and barriers and of carving up the earth into warring and competing nations and factions. Furthermore, "Thus founded on the severance of nature's bonds, thus sustained by the suspended sword, thus outlined in blood still crying for vengeance, the geography of conquest, like the quicksands of the ocean, is ever shifting its frontier, ever subject to the inroads of avarice and ambition" (Brewster 77). To return to the poem's final lines, we can notice the ways that Whitman replaces the tension-fraught nature of human proprietorship over the land with a fundamental, and democratic, form of divine ownership. In the poet's final invocation, "O my brave soul! / O farther, farther sail" (252–53), we see that the territory traversed is owned by the divine in ways that guarantee safety and security that "human-owned" spaces might need due to nationalism or imperialist greed: "O daring joy, but safe! are they not all the seas of God?" (254). If the invocation of the phrase "passage to India" had become synonymous in nineteenth-century America with manifest destiny, Whitman transforms it to mean a cosmologically deeper, divine, and "trackless" expanse that floats free of America's real or imagined borders.

Whitman's self-conscious spiritual quest, his attempts at poetic embodiment, and his embrace of emotional excess and abstraction do not easily align with T. S. Eliot's theories of impersonality or Ezra Pound's imagist ideals of poetic hygiene. But even if these qualities necessitated some creation of distance on the part of early twentieth-century poetic innovators, Whitman's experiments with form and content made him an honorary, anachronistic, or early modernist for many writers in the United States and beyond.[21] Louis Untermeyer's preface to his 1919 *Modern American Poetry* considers Whitman a part of the contours of the new poetic tradition that his anthology seeks to map, calling him "the greatest of the moderns who showed the grandeur of simplicity, the rich poetry of everyday" (vii). Sylvia Beach and Adrienne Monnier's Parisian bookstore, Shakespeare and Company, displayed a photograph of Whitman on the wall and also put out for display several small Whitman manuscripts that had belonged to Beach's Aunt Agnes Orbison, who herself had visited Walt Whitman in Camden (Beach 20). Langston Hughes, Gertrude Stein, and H.D. shared this sense of Whitman's importance. Stein did draw playful attention to the excessive qualities of Whitman's work in section 52 of her poem "Yet Dish": "Leaves of gas, leaves of get a towel louder" (*Yale Gertrude Stein* 61). But despite any possible dyspeptic or windbag qualities of *Leaves of Grass*, or the notion that someone, perhaps a poet like Stein, might have to wipe up after Whitman or shout over him, Stein grants him unquestioned status as a twentieth-century modernist who lived before his time: "And the United States had the first instance of what I call Twentieth Century writing. You saw it first in Walt Whitman. He was the beginning of the movement" (*How Writing Is Written* 153). H.D.'s 1937 book review of Edgar Lee Masters's biography, *Whitman*, provides a more extended consideration of Whitman's importance. The Whitmanian title H.D. chose for her review, "I Sing Democracy," signals her claims that Whitman shaped not only the American literary tradition but America itself and all in its purview. If Abraham Lincoln created a war, then "Walt Whitman created a continent" (159).[22] These democratic qualities made him one of Langston Hughes's most valued influences, inspiring Hughes to develop his own version of anaphoric parataxis. Hughes, like Whitman, uses this poetic strategy to grapple with the major geographical epistemologies of the day, in his case environmental determinism—geography's next major phase of development in American culture.

3

African Diasporic Re-Placing

Race and Environment in the Poetry of Helene Johnson and Langston Hughes

Hughes's 1921 poem "The Negro Speaks of Rivers" seeks a version of the "vast terraqueous globe" that Whitman pursued, its routing/rooting in river systems driven by a particular investment in African diasporic and African American sensibilities. The similarities between this poem and Whitman's "Passage to India" in creating linkages through content and poetic form among disparate global geographies are reflective of the broader continuities we can draw between Whitman and the modernist period. Hughes's depth of affinity for "Old Walt," to cite the title of his 1954 poem about the poet, is little surprise.[1] In "Calls Whitman Negroes' First Great Poetic Friend, Lincoln of Letters," one of several *Chicago Defender* columns Hughes published on Whitman, Hughes describes him as "the greatest of American poets" not only because of his pioneering commitment to "the basic precepts of American democracy as applying to everyone, white or black," but also because of his innovations as "the great pioneer of 'free verse' in America, taking his rhythms from the rolling sonorities of human speech and the majestic poetry of the Bible" (11).[2] As George Hutchinson has observed, Hughes was not merely working within the Whitmanian tradition but actively responding to his explicit invitation in generating an "African American–based poetic syncretism" ("Langston" 20). Whitman was in this way a notable exception to the canonical, Anglo-American tradition of literature and learning from which Hughes felt the need to distance himself upon the start of his maritime adventures in his early twenties. In the opening of volume 1 of his autobiography, *The Big Sea,* he recounts the grand gesture of throwing over the rail of the SS *Malone* "all the books I had had at Columbia, and all the books I had lately bought to read," liberating himself at the start of his first journey to Africa of literature that was like "a million bricks" weighing down his heart (3). In a draft version of the text, he had planned to say he had tossed everything but kept *Leaves of Grass,* positioning this seminal work by Whitman as a category apart from the literature that he felt was oppressive (Rampersad, *The Life* 414n72).[3]

Hughes's "The Negro Speaks of Rivers," the central focus of this chapter, evokes this sense of connection to Whitman on a number of levels. We can note in this poem, as George Hutchinson has done, a "triply descended 'I'" that looks back to the blues, to spirituals, and to Whitman's inclusive and absorptive "I" ("Langston" 21). But we can also find in both "Passage to India" and "The Negro Speaks of Rivers" a "terraqueous globe" in which globality springs from water systems—for Whitman, the ocean, and for Hughes, rivers. More important, both writers share a common engagement with major geographic discourses of the moment and innovative efforts to use poetic form, particularly anaphoric parataxis, as a space-collapsing technique that advances a democratic reconfiguring of global geography. While Whitman's cosmographical oneness presents an alternative to the nation, Hughes's geographical consolidation of African diasporic geography challenges the seemingly fixed and fixative ideas about both nation and race.

Unlike the modernist tradition embodied by Whitman, African American considerations of the environment and the landscape have been understandably complicated by multiple historical legacies that overdetermine African American and African diasporic connections with the natural environment: forcible removal from Africa and then forced agricultural labor in the Americas under slavery, land dispossession in Africa under colonialism, and the often rural, wooded settings of lynchings. In her preface to her powerful anthology of poetry, *Black Nature,* Camille Dungy notes the wide variety of poetic strategies and responses to nature, generated partly because of this "manner in which the natural world has been used to destroy, damage, or subjugate African Americans" in a larger "active history of betrayal and danger in the outdoors" (xxvi). Hughes's 1924 poem "Lament for Dark Peoples" registers the impact of being severed from a beloved African environment by the advent of slavery: "They drove me out of the forest. / They took me away from the jungles. / I lost my trees. / I lost my silver moons" (*CP1* 39).[4] The speaker's factual, declarative statements outline a loss of beauty that contrasts with his new reality, imprisonment in a West that is not as orderly and rational as it purports to be but instead is merely an excessive and even chaotic performance, a "circus of civilization" (*CP1* 39).

This kind of critique of western civilization specifically mourning the loss of African land, whether through slavery or colonization, animates a number of African American modernist lyrics of the 1920s. In Anne Spencer's "White Things," Africa appears like it does in Hughes's "Lament" as a "silvered world"; here, Africa's beauty is also figured as vibrantly colorful, whose "earth-plains fair plains" are both bucolic and democratically free and whose skies are "golden stars with lances fine" and hills that are "all red and darkened pine." These features become "blanched" by the exercise of white colonial dominance, which Spencer links to a second kind of whitewashing violence in which black bodies become "ashes white" by the flames of lynching (Dungy 155). We see a similar frustration with the geographical manifestations of social and political exertions of power in Helene Johnson's poem, "Bottled." The conclusion to the poem likens a bottle containing African sand to an African American street musician dancing in New York whose performance style is similarly contained, only in this case by the disapproving reactions of urban onlookers who hew to a narrow sense of what is acceptable: "Say! That man that took that sand from the Sahara desert / And put it in a little bottle on a shelf in the library, / That's what they done to this shine, ain't it? Bottled him" (Honey 98). The sense of African soil or sand as out of place (on a shelf in a library exhibit in New York), removed from its organic context and contained by presumably white machinery, haunts the poem as a trauma of the African diasporic experience.[5]

Environmental determinist discourse introduced yet another layer of complication, adding to the many negative historical associations attached to the natural world for African Americans. Not only did it read African American and African diasporic identity through the influencing features of a tropical African environment that for the most part was not territorially possessed and administered by African peoples themselves, but it did so through a logic that reduced this environmental cradle to one that produced cultural depravity or simplicity as its human result. It is by now well understood that the New Negro aesthetic of the 1920s and early 1930s conjured forth Africa as a romanticized image of cultural and racial pride. But it is important to recognize the geographic dimensions of this particular kind of romanticized image, which necessarily intersects with dominant spatial epistemologies in academic and popular culture in which environmental determinist formulas about land and people circulated. For poets such as Anne Spencer, Helene Johnson, and Langston Hughes, the representation of African landscapes alters the terms on which environmental determinism underwrote arguments about race, culture, and nation.

By the time this group of poets published their poems of the 1920s, environmental determinism had become firmly entrenched in both academic and popular geography. These sources of geographic knowledge circulated the same racial hierarchies generated by nineteenth-century Social Darwinist theories, although attributing reasons for the presumed inferiority of nonwestern races to the inhibiting effects of physical environment or to a combination of an inhospitable environment and the inability of local peoples to properly exploit whatever material advantages their regions presented. The overall force of environmental determinism's claims about nonwestern peoples as well as the various ways that social and governmental entities used those notions against the non-West thus posed a direct threat to African American writers looking to find meaning in their African heritage. My interest here is in how African American poets reframed and rejected the sorts of thinking about race that geography endorsed in both academic and popular forms. "The Negro Speaks of Rivers" in particular seizes the authority of geography as a knowledge project while challenging environmental determinist ideas in several important ways.

One of the most powerful aspects of the poem long recognized by scholars is the way that it positions the speaker as a knowing subject, whose experience and wisdom counters racial discourses that would deny intelligence and depth to the African American subject. The speaker asserts his claims of knowledge three times in the poem, twice in the first two lines alone: "I've known rivers: / I've known rivers ancient as the world and older than the flow of human blood in human veins" (*CP1* 23). The object of the speaker's knowledge is, of course, slices of time past that are connected to specific geographic environments. The relay between the human and the landscape in the poem enters the geographic conversations of Hughes's day, overturning the fundamental cause-and-effect logic of environmental determinism that sustained primitivist stereotypes about race. With his rivers that "lull" and "sing," and his speaker whose soul, body, and mind have deepened with the ages, Hughes replaces the crude metrics of anthropometry with the suggestive richness of anthropomorphism. Hughes also undermines the ways that environmental determinism supported conventional, dominant narratives of American development that systematically excluded and suppressed race and with it American exceptionalist ideas that needed to define African geography and the people connected to it as inferior.

In this sense, I take the title of Hughes's poem at its word: the African American speaker of "The Negro Speaks of Rivers" expressly performs the function so familiarly carried out by the geographer who "speaks" about waterways and other topographical features as he or she attempts to make sense of the world. The title signals the obvious way that the speaker confidently occupies the seat of knowledge production typically assumed by white western writers of geography and which generated racist definitions of nonwestern people as seen through, and limited by, their environments.

First and foremost, by emphasizing in his title that he is "speaking" of rivers instead of "writing" about them, Hughes indigenizes the operations of academic and popular geography by recasting academic and magazine writing conventions within a distinctly oral tradition. The knowledge conveyed in the poem joins a long history of oral tradition in African and African American culture, and this transformation thereby subsumes western academic and popular geographic knowledge production into a key cultural and artistic form of the African diaspora. Whatever follows in the poem proceeds from an oral cultural articulation that the speaker controls, rather than the institutional and commercial elements of geography produced by white dominant culture.[6] It is neither the typical academic geographer nor the *National Geographic Magazine* speaking of rivers here.

At the same time that the poem seizes the authority of the geographer-writer, now geographer-speaker, it changes the perspective and scale of academic geography: the geographer-speaker operates both as observer and as an active participant in the poem. By casting the poem in the singular "I" and gesturing only implicitly toward entire communities that the "I" might represent, Hughes reformulates academic and popular geography's tendency to generalize across entire communities and regions. Its particulars, based in personal, experiential knowledge, counterpose institutional knowledge based in generalizations and interrupt the geographical matrices attached to race. The anaphoric repetition of "I've known" is an obvious rejection of primitivist stereotypes that would deny the African American subject intellectual capacities, restoring personhood to an African American speaker who not just "knows" but knows a lot, the repeated phrase allowing the knowledge to accrue over the course of the poem. But it also reinforces the value of personal experience of geographical and historical sites of significance or the value of belonging to the social groups connected to the regions in question.

This reading of "The Negro Speaks of Rivers" as engaging with mainstream geographic notions aligns with scholars who have recently shifted away from assumptions that the poem is an immediate and largely unmediated composition, ideas about the poem that Hughes himself encouraged. According to his often-cited account in *The Big Sea* of the origin of the poem, Hughes's vision of the Mississippi River from his train window while he traveled to Mexico in 1920 to visit his father brought this poem into being in a flash of inspiration and transcription. The river viewed from the moving train, Hughes recounts, led him "to think about other rivers in our past—the Congo, and the Niger, and the Nile in Africa—and the thought came to me: 'I've known rivers,' and I put it down on the back of an envelope I had in my pocket, and within the space of ten or fifteen minutes, as the train gathered speed in the dusk, I had written this poem" (*Big Sea* 55). The primacy of this account has been challenged by Ira Dworkin, who notes discrepancies between the published version and Hughes's later account in *The Big Sea* of the poem's contents. The "internationalist imperative" Dworkin locates in the poem reflects, instead, a much wider array of mediating effects of contemporary world events that were of interest to the African American community in the late 1910s (633).[7] The poem is indeed more complicated than its famous origin story might suggest, and recognizing the mediated nature of its ideas enables us to gain more purchase on its terms of engagement with dominant paradigms about geography and culture in the early twentieth century.

Considering Hughes in relation to these ideas may run counter to critical tendencies in part because of the ways that African American writers visibly aligned themselves with new anthropological developments, which partly overlapped with but were highly critical of the foundational assumptions of environmental determinist academic geography. Franz Boas, today well known in African American literary studies for his role in training Zora Neale Hurston at Columbia and Barnard, who famously used the lens of anthropology to examine African American folklore (Sollors 165),[8] criticized as early as 1894 reigning anthropological discourse for its racist classifications of groups based on biological race (L. Baker, *From Savage* 104). His criticism of geography's environmental determinist principles was formulated even earlier, growing out of the minor in geography that formed a part of his graduate training (101). In his 1887 essay "The Study of Geography," Boas begins to formulate the limitations he sees in the dominant methodological operations of anthropogeography, which he characterizes as a focus on "the discovery of general laws" gleaned from phenomena. Instead, he argues for the importance of a sort of geography that would focus all its energies on "the investigation of phenomena themselves" in a particular culture and their organic evolution (175). His 1911 *The Mind of Primitive Man* voiced a fuller repudiation of environmentalist principles and launched a full-on assault.[9] At the same time, this new anthropology directly enabled African American writers, artists, and intellectuals to reclaim their African heritage. Boas's systematic academic attention to sub-Saharan Africa in particular, as opposed to northern regions in Africa or other non-European locations around the world, encouraged a new sense of legitimacy in perceptions of the populations and cultures of Africa. The larger perspective that his scholarship took claimed that culture was not reducible to biology and that nonwestern cultures needed to be understood in relation to their own internal development, which contributed significantly to anthropology's shift away from the Social Darwinist ideas of the nineteenth century (L. Baker, *Savage to Negro* 5). Boas vocally supported Africanist and African American projects by participating in the first Pan-African Congress and the Second National Negro Conference and by serving for all intents and purposes, as George Hutchinson puts it, as "the house anthropologist for *Crisis* magazine" *(Harlem Renaissance* 63). For these reasons and others, anthropology has been the interdisciplinary link of choice among scholars who have focused on African American epistemological and institutional engagements.

African American intellectuals, writers, and artists would have encountered environmental determinism in several ways, however, and with it geographically formulated racism that was only slightly more attenuated than the blatant Social Darwinist theories about race. Although it aimed to shift the focus away from biologized ideas about racial and cultural difference, environmental determinism nevertheless supported racial hierarchies, albeit based on environmental influence rather than physical difference per se. In this way geographers contributed to an "anthropometric" project, as David Livingstone calls it, which measured ethnic groups against each other and attributed ethnic character directly to regional climate (225). Environmental determinists working in this vein generated an "ethnic moral topography" of the globe that colluded with sociologists and those anthropologists unlike Boas who explicitly endorsed racial classifications that relegated African identity to the bottom of the scale (221).

These ideas constituted a racial politics through geographical means, and

African Americans knew of them in both their academic and popular forms. Ellen Churchill Semple's 1911 *Influences of Geographic Environment* was reviewed extensively in the *Nation,* a magazine that many African Americans, including Hughes, read regularly, and it also appeared in a list of notable books in the *Crisis* in the same year it was published.[10] The *National Geographic Magazine* published countless articles pursuing an environmentalist angle in its constructions of Africa, coverage of which the *Crisis* periodically highlighted. A. W. Greely's "Recent Geographic Advances, Especially in Africa," which appeared in the April 1911 issue of *National Geographic Magazine,* for instance, appears among the titles listed in the "What to Read" section of the 1911 *Crisis.*[11]

One of the better-known purveyors of the "moral topography" was Semple's contemporary Ellsworth Huntington, who focused on the shaping force of climate. He advanced his ideas most visibly in the 1924 *Character of Races,* published several years after Hughes's poem was written. Anthropometric geography, with its long-standing history, animated a number of environmental determinists, but none so forcefully and consistently as Huntington. His 1919 *World-Power and Evolution,* for instance, distinguished between variable weather (i.e., the temperate zones of the West) that encourages stimulation and civilizational development (the development of European and American cultures) and the monotony of more extreme climates (the Arctic and the tropics) that minimizes stimulation and lowers activity levels (such as for Eskimo and African peoples). A review of *World-Power and Evolution* used these ideas as explanations for the ostensible relative ease of integration among Jews in the United States compared with people of African origin: "organic variability, which is adaptability to circumstance and climate, becomes the clue to the superiority of the Jew, concretely contrasted with the Negro, as both have been transferred to our continent and mode of life" ("Literature: Mental Reconstruction" 871). In this example, the Jewish people are held up for praise while people of African descent are denigrated because of the formative benefits of the temperate climate of Europe availed by the former group but denied to the latter.

Even Semple's *Influences of Geographic Environment,* which explicitly declared its intention to shift away from theories of race, dips into these anthropometric waters. When discussing what happens when Europeans acclimatize to the tropical zone of the so-called darker races, she warns of "derangements in the physiological functions of heart, liver, kidneys, and organs of reproduction" because of the "intense enervation" that takes hold in white settlers in the tropics (Semple, *Influences* 626). Semple's characterization could just as easily be describing the problems of physical and psychological derangement that befall Joseph Conrad's hapless European characters, Kayerts and Carlier, when they arrive at their new post in the Belgian Congo in "An Outpost of Progress," a story that draws from the same image bank and plot structure as *Heart of Darkness*. For Semple, because such forms of racial conditioning and adaptation evolve over vast expanses of time, "the conquering white race of the Temperate Zone is to be excluded by adverse climatic conditions from the productive but undeveloped Tropics, unless it consents to hybridization" (*Influences* 628).[12] The only immediate solution for white Europeans is race-mixing as a shortcut to the gradual acclimatizing process, and while Semple uses the term *hybridization* rather than a more freighted term from the period, like *miscegenation*, her comment shares the assumption that hot climates produce races that are fundamentally uncivilized and incapable of progress.

These ideas did not merely circulate in academic circles of professional geographers. They appeared in the American public school system and again, in different form, in the *National Geographic Magazine*. American curricular reform at the turn of the twentieth century aimed to revitalize geography as a subject by drawing environmental determinist geography into textbooks and into the classroom (Schulten 122). Semple joined many other prominent academic geographers advocating for this shift and described it as an improvement over textbooks that had tended to contain vast quantities of facts that were "stultifying" to children. Causal ideas regarding the impacts of environment on human societies, by contrast, are ones "the child will seize upon and retain" (Semple, "Emphasis upon Anthropo-Geography in Schools" 208). Such equations found their way into primary and secondary school textbooks and with them an affirmation of racial hierarchies, as we see in a 1901 Rand McNally geography textbook: "Most of the civilized people of the world belong to the white race, though in some countries the people of that race are half-civilized. The savages belong to the red, brown, and black races. Most people of the yellow race are half-civilized, but you will read some day of the yellow people of Japan. They are the only great people of that race that has become civilized" (qtd. in Schulten 115). Alongside such explicit formulations were forms of environmental essentialism that expressed racial hierarchies in slightly less explicit ways. Dodge's high school–level *Comparative Geography*, for example, also published by Rand McNally, proclaims the Germans to be "naturally home-loving and quiet, painstaking and thorough;

they are less impetuous than the French." In France's "large cities . . . and especially in Paris, the people are very polished and fond of excitement. The latter characteristic is expressed in their manner of speech, in their movements, and in their fondness for political struggles. In rural communities, on the contrary, the people are more slow of speech and action, more satisfied with the monotony of their daily life, and less easily excited" (qtd. in Rothenberg 81n56).

These ideas were reinforced in the *National Geographic Magazine*, whose circulation had grown from 337,000 in 1914 to surpassing 625,000 in 1917 due in part to its coverage of Allied nations and their sacrifices in World War I (Rothenberg 64–65). By 1919, its circulation was overshadowed by the *Saturday Evening Post*, but that figure exceeded the combined circulation of all of these magazines put together: *Atlantic Monthly*, the *Century*, *Harper's Magazine*, *Outlook*, *Review of Reviews*, *Scribner's*, and *World's Work* (Schulten 153). As I document in chapter 1, Hughes professed to have loved the *National Geographic Magazine* his entire life, although this proclamation of love does not necessarily indicate an uncomplicated relationship to the magazine's content nor to its policies. It is useful to consider that Hughes declares his affinity for the magazine but not a lifelong membership in the National Geographic Society. In the early decades of the twentieth century, the magazine was not sold at newsstands or distributed through purely commercial subscription; society membership was the only route to an individual magazine subscription. While occasional exceptions were made in granting memberships to people of color, as in the case of Ethiopian emperor Haile Selassie, the society's president and magazine's editor-in-chief, Gilbert H. Grosvenor, measured its ideal number of projected memberships for "every community of 100 white people in the United States" (qtd. in Poole 63). Such race-exclusionary membership policies were visible in the organization's official policies until well after World War II (Poole 63). While this aspect is important to note, what is also true is that the realities of the readership extended beyond the white boundaries the magazine may have set out for itself and included African Americans who, as Stephanie Hawkins has documented, were among those who wrote to the magazine complaining about its distorted depictions of African peoples (27, 149).

It is admittedly difficult to pinpoint Hughes's familiarity with particular issues of the magazine, though given the formulaic nature of *National Geographic Magazine*, this may not be as much of a problem for scholars as it might be with other sorts of periodicals. Constructing a timeline of engagement for a single reader may, in fact, be especially problematic with magazines like *National Geographic,* because individual issues enjoy a kind of longevity that other magazines can only hope for, forming a sort of reference library for subscribers who might read volumes repeatedly or out of order. The National Geographic Society's certificate of membership and the middlebrow educative mission of its magazine, as Hawkins has observed, meant that its readers "treated the magazine not only as a kind of geography textbook, but also a field guide, a portable university, even a classic of literature and art" (31). Individual issues were retained well past their publication date rather than discarded also because *National Geographic Magazine* very quickly came to serve as an emblem of educated, refined taste (not to mention the fact that it was a fairly expensive magazine compared with other periodicals).[13] Many of its core ideas were modulated if not repeated outright across multiple issues. Perhaps more so than many other periodicals, then, the *National Geographic Magazine* can be said to convey to its readers an accretive set of stock ideas that are not restricted to the boundaries of specific temporal windows of reading.

It is hard to imagine that the *National Geographic Magazine*'s repeated images of both American triumphalism and African inferiority would have escaped Hughes's notice as an avid reader. While it basically ignored the presence of African Americans within the United States throughout the first few decades of the twentieth century, numerous articles featuring African locales perpetuated the kinds of primitivist ideas about racial otherness that we might find in the modernist primitivist visual art of Picasso and Matisse, only with a much more widespread audience to consume them. Many of the early issues of the magazine conveyed the notion that while America both benefited from and also superseded the gifts of its temperate climate, the hot African climate produced lassitude and passivity among people residing there.

Gardiner Greene Hubbard, then president of the National Geographic Society, outlined these environmental formulas of racial character in his 1897 essay in the *National Geographic Magazine* concerning the "the effects of geographic environment in developing the civilization of the world" ("Synopsis of a Course of Lectures" 29). Describing the subject selected for the popular lecture series the Society would run that year, Hubbard notes that while environment was the primary cause of development in the early stages of human history and continues to exert influence, "as man advanced in knowledge and intelligence he became more and more independent of his surroundings" (30). This resulted, ultimately, in a new formulation of the humanity and landscape equation describing the New World, "where the environments become subservient to man and not man to his environments" ("Synopsis" 32). His assertion suggests one of the central ways that environmental determinism supported ideas of western superiority and American exceptionalism, given its assumption that the West is able to "evolve" out of environmental determinism whereas the rest of the world is still at the mercy of local pressures. In fact, in outlining the projected focus on America of the ninth lecture in the series, he claims, "All geographic environments have become subservient to the will of the people, from ocean to ocean, from the waters of the Hudson to the waters of the gulf of Mexico, one people and one language, an American race, an empire vaster than that of Rome, home of all the nations of the world, welded into one great and free people" (Hubbard "Synopsis" 32). Those who can conquer the environment will triumph, as in America, while those who cannot, as in nonwestern regions such as Africa, are doomed to a sort of primal thralldom to the land.

The condescending, negative view of the qualities of African peoples takes an overt form in Hubbard's essay "Africa, Its Past and Future" (1889). Although he notes the devastation wreaked in Africa by the European slave trade as a form of exploitation, almost in the same breath he declares that the "temper and disposition of the Negro make him a most useful slave. He can endure continuous hard labor, live on little, has a cheerful disposition, and rarely rises against his master" ("Africa" 112). African subduability seemingly stems from environmental context: "Nature has spread a bountiful and never-ending harvest before the Negro, and given to him a climate where neither labor of body or mind, neither clothing nor a house, is essential to his comfort. All nature invites to an idle life; and it is only through compulsion, and contact with a life from without, that his condition can be improved" (Hubbard, "Africa" 123).

These early essays from the society's president dictated the schema, so to speak, within which environmentalist perceptions of Africa would be expressed in numerous other *National Geographic Magazine* essays, particularly those published closer to Hughes's composition of "The Negro Speaks of Rivers." This is not surprising given that Huntington, the anthropometric environmentalist *par excellence*, published seven articles in the *National Geographic Magazine* between 1908 and 1910. In another example, Edwin Grosvenor's 1918 article "Races of Europe" rejects biologized explanations for

Europe's superiority in the ways that academic geographers like Semple do: "That the early ancestors of the present European peoples were more highly endowed than their kindred or contemporaries or possessed greater capacity for development, there is no reason to believe" (443). Instead, it is the "physical advantages" of the environment that have given Europeans their "unquestioned superiority" (443). Although much of the essay taxonomizes the characteristics of various European races as understood in relation to their particular region's distinctive features, the introductory frame of the essay declares the "progressive" nature of all of Europe's races, no matter their particular inflection. This Grosvenor attributes to the fact that practically all of Europe lies in the northern half of the "North Temperate Zone," which frees Europe of the limitations imposed by the earth's more extreme arctic or tropic zones:

> Nowhere, except in the farthest boreal limits, does excessive cold stunt body and mind. Nowhere does excessive and continued heat sap energy and enervate the will. No spontaneous prodigality of Nature removes the necessity of exertion and induces sloth. (443)

Not only is the continent of Africa implicated among those sloth-producing environments but it is monolithically so because it is set apart from Asia by a distinct isthmus and lacks the stimulating effects that Europe enjoys due to its shared eastern border with Asia.

A number of poems by African American writers during the 1920s can be seen as participating in this larger conversation about the nature of African environmental influence and recasting it as a consideration of the degree to which the environment allows for cultural or even individual artistic agency. Helene Johnson's 1926 poem "Magula" reveals environmental determinism to be potentially as dangerous as the Christianizing push of European colonization in Africa. The poem stages a dramatic monologue by an unnamed speaker who encounters Magula, a woman indigenous to an African locale. She is listening to a religious and presumably western white man carrying a Bible and who wants to convert her to a religion that would prohibit dancing and, along with it, other aspects of her indigenous African traditions and values.[14] The African environment in which the speaker finds Magula is a "pulsing, riotous gasp of color" formed by colorful flamingoes, dark mangroves, and green lizards and crocodiles (Hughes and Bontemps

152). The speaker promises her poetry instead of scripture and warns her of the restrictions that the man's western religion would bring with it.[15] She asks Magula if she would sell her colorful sunset, her fragrant flowers, and "the passionate wonder of your forest" for a religion "that will not let you dance" (153). The speaker's own gender and position are left unclear (many critics infer the speaker to be female, like Johnson), although her invocation of seraphim and poetry suggests she is a westerner (Lynes, "'Sprung from American Soil'" 534). The speaker's characterization of what Magula would lose upon falling under the spell of western Christianity threads the needle between a western perspective less enlightened about the nature of primitivist and environmental determinist moves and a presumed African diasporic position of sympathy. On the one hand, the speaker's constructions of Africa wear the patina of environmental determinist formulas that would view the excesses of the African environment as producing laziness: "Do not let him lure you from your laughing waters, / Lulling lakes, lissome winds."[16] This presumed lassitude is, however, potentially countered by "the passionate wonder of your forest," which could suggest that the physical environment promotes an invigorating curiosity among its inhabitants. But when "wonder" is attached to "passionate," the image suggests more of a childlike sense of emotionally excited awe than intellectual or creative contemplation of larger things. The discomfort the poem produces about exactly how the speaker, and Magula, are positioned in relation to environmental determinism is precisely the point regarding this discourse's malleable nature and sometimes subtle operations of power.

A similar desire to revalue the influence of the African environment and a similar presentation of alternatives to environmental determinism's tactics can be located in a significant number of Hughes's poems gathered in his 1926 collection, *The Weary Blues,* where "The Negro Speaks of Rivers" also appeared after its initial publication in the *Crisis.* "Our Land" operates as a wish fulfillment for the environment of Africa. Multiple registers of meaning in the title suggest that the West is, tragically, the land that the speaker actually has to consider as his own, and that Africa, a truer or longed-for form of possession, is lost to him; reconnection to it can only be conjured up through the poem. Hughes describes an ideal landscape in a way that operates like an itemized list of the environmental features of the tropical "extreme" zone of Africa, replete with sun, water, and a colorful landscape of parrots and colored bandanas.

> We should have a land of sun,
> Of gorgeous sun,
> And a land of fragrant water
> Where the twilight
> Is a soft bandanna handkerchief
> Of rose and gold,
> And not this land where life is cold. (*CP1* 32)

This first stanza's reversal of typical environmentalist dismissals of the tropical zone nears completion when it portrays the ostensibly "temperate" zone of America, "this land," in a negative light as a cold environment that lacks beauty and softness. The anaphoric repetition of "we should have" implies that these physical features help to produce the human products and experiences that take place within: "Ah, we should have a land of joy, / Of love and joy and wine and song. / And not this land where joy is wrong" (*CP1* 33). The poem emphasizes love, joy, wine, and song as things "of" the land that are created, cultivated, civilized. The "song/wrong" rhyme carries an argument that music and artistry are more supported by Africa than they are by the United States.

If "Our Land" longs for connection with a tropical African environment in a way that reverses its moral topography or moral coding, "Poem [1]" imagines a speaker who fully retains a connection to Africa. Here, the features of the natural environment directly impress themselves upon the speaker such that "All the wild hot moons of the jungles shine in my soul" (*CP1* 32). The adjectival pileup of the "wild hot moons" conflates two parts of the environmental determinist equation: that extreme heat ("hot") produces unrestrained and uncultivated societies ("wild"). Although these external environmental features penetrate the speaker's soul, they "shine in my soul" in a way that is celebrated rather than morally denigrated. This revaluing participates in a larger reversal of the entire environmental determinist moral topography, such that western civilization is not superior because it possesses a climate felicitous to civilization; rather, it is to be feared precisely because it is cold:

> I am afraid of this civilization—
> So hard,
> So strong,
> So cold. (*CP1* 32)

Normally positive qualities that the West celebrates about itself, hardness and strength, are associated with coldness, which gives it a stunted and unfeeling emotional valance that reveals the problematic underbelly of the West. Given that this poem appeared next to "Our Land" in *The Weary Blues,* the "cold" of this poem implicitly evokes the lack of beauty, protection, and aesthetic decoration associated with the West in "Our Land." There, a tropical climate where "the twilight / Is a soft bandanna handkerchief / Of rose and gold" contrasts with "And not this land where life is cold" (*CP1* 32). If we treat "so cold" as its own syntactic unit, as Hughes's line breaks and indentations invite us to do, it also revises the subjective interpretation of climate itself that environmental determinists following in the footsteps of Huntington make a part of their dominant narrative. For them, Europe and much of America occupies not a cold but a "temperate" zone. But Hughes's barometer is calibrated to use the tropical climate as his norm. Significantly, this is something that Hughes himself cannot realize, and so he has to project these connections to the African tropical landscape into the ekphrastic conceit of the poem's subtitle, "For the portrait of an African boy after the manner of Gauguin" (*CP1* 32).

"The Negro Speaks of Rivers" also tangles with some of environmental determinism's fundamental ideas, dismantling the logic and content of the sorts of pronouncements made by Grosvenor, Huntington, and Hubbard in a way that recasts the causal links between landscape and cultural and racial identity to grant the African diasporic subject agency and significance. The poem's formal qualities, however, deliberately use anaphoric parataxis to place side by side not only temporal fragments but also geographical ones from various parts of the African diasporic world. In the process, the poem both mirrors the moves of environmental determinism's global comparison and jams the machinery of some of its oppressive conclusions about race and culture.

Hughes devotes the entire first stanza to the rivers his poem examines in ways that echo the environmentalist style of the *National Geographic Magazine* that places landscape first, people second: "I've known rivers: / I've known rivers ancient as the world and older than the flow of human blood in human veins." Here, Hughes highlights in a quite literal sense the ways in which the world's rivers came into being first, before the earth's inhabitants, which grants the landscape a kind of developmental firstness that environmental determinists would stress. In several of the images that follow, the speaker is nurtured by various aspects of the landscape. We might think most immediately of Semple's image of the earth mothering its inhabitants in the opening pages of her 1911 *Influences* discussed in chapter 1. Its imagery resonates with the nurturing qualities the poem gives to the Euphrates and Congo Rivers: "I bathed in the Euphrates when the dawns were young," suggesting the key role the river played in the early development of the human race. These life-giving and civilization-giving qualities of the Euphrates are featured prominently in James Baikie's 1916 *National Geographic Magazine* feature article on the Euphrates, appropriately entitled "The Cradle of Civilization." The Congo performs a similar function in providing the speaker with rest and recuperation: "I built my hut near the Congo and it lulled me to sleep." Here, the image runs very close to the primitivist ideas of environmentally produced lassitude and laziness constructed by environmental determinist and *National Geographic* formulations: the river literally induces sleep in the speaker. The key difference, however, lies in the fact that this image immediately follows one of work and of self-sustenance in the form of the speaker having built his own hut. Sleep in this case is earned sleep.

As specific geographic vignettes unfold, they reverse in bold ways the key aspect of environmental causality central to Semple, Huntington, and their orbit of influence. Most tellingly, the poem projects human qualities onto the rivers ("I've known rivers: ancient, dusky rivers") and thus conflates the natural environment with the human in a masterful act of poetic compression. The racial modifier "dusky" is a familiar one in eighteenth- and nineteenth-century racial description, particularly within abolitionist poetry, and we also see it used in Semple's *Influences* when it refers to the Native American women whom European fur traders often married on the frontier as "dusky squaws" (47). Here, this racially weighted term modifies the natural environment and not the other way around. The anthropometric becomes anthropomorphic. At the same time that Hughes suggests correspondences, he is careful to preserve boundaries between subject and environment. The repeated refrain "My soul has grown deep like the rivers" does, in one sense, imply that the speaker's experience of each river and its history has struck his soul, affecting him incrementally or cumulatively. However, Hughes's simile sets up a protective distance where metaphor would not, such that he can focus on the significance of the environment as dominant academic and middlebrow geography does but in order to convey something that ultimately originates in the human subject.

This use of simile to unite and divide is consistent with the poem's overall pattern of anaphoric parataxis, the repeated "I"-plus-verb construction in the middle section of the poem which links individual vignettes through repetition and association rather than through explicit forms of linguistic subordination. This space-collapsing poetic form brings multiple major locations within the African diaspora into one spatial orbit on the space of the page and through the experiences of the speaker's body. These geographical elements cycle through ideas of birth, death, bondage, and freedom: the Congo River, connected from a diasporic point of view with origins and dispossession, is, like the Euphrates, the "cradle of civilization" with its biblical resonances of paradise lost. (Genesis lists the Euphrates as one of Eden's four rivers.) By contrast, the Nile, with its pyramids monumentalizing death, conjures up captivity as well as the promise of exodus and liberation, as does the flood-prone Mississippi, down which the youth Lincoln traveled to New Orleans, its slave market, and his own revelation about slavery's gross injustices. (In the light of sunset, the Mississippi's "muddy bosom" can turn "golden.") In "The Negro Speaks of Rivers," four rivers speak to one another and exist alongside each other spatially, through the logic of a shared diasporic experience, freed of their conventional positions in time and space.[17] Hughes creates a diasporic heterotopia, to use Michel Foucault's concept, in which normally unlike things are brought together into a new configuration; such heterotopias can "suspect, neutralize, or invert" the conditions they "designate, mirror, or reflect" (24). Conventional geography as well as the discourses used to represent its relationship to race and culture are countered here; in Hughes's configuration, the speaking subject and the environment are bearers of human suffering and aspiration as well as the coordinates of global civilizational significance.

The poem thus privileges affiliations that register and elevate the African diasporic attachments of the speaker that exist alongside or even run counter to a singular identification the speaker might make with America and its history. In this regard, the poem's river names and imagery generate a specific challenge to national narratives of growth and expansion that fuel American exceptionalism. Environmental determinists devoted considerable attention to river systems as a natural resource and practically a necessary precondition for the development of a civilized society, as Semple notes in *Influences*:

> Rivers tend always to be centers of population, not outskirts or perimeters. They offer advantages that have always attracted settlement—fertile alluvial soil, a nearby water supply, command of a natural highway for intercourse with neighbors and access to markets. Among civilized peoples fluvial settlements have been the nuclei of broad states, passing rapidly through an embryonic development to a maturity in which the

> old center can still be distinguished by a greater density of population. Only among savages or among civilized people who have temporarily reverted to primitive conditions in virgin colonial lands, do we find genuine riverine folk, whose existence is closely restricted to their bordering streams. (363)

"Civilized" western societies evolve from "embryonic development" to "maturity" because of what their river systems make available to them, in contrast to the "savages" who fail to generate a dense population and to exploit the qualities of trade and exchange that river systems allow, a notion that Semple naturalizes through her petri-dish conceit.

America's success, for Semple, hinges upon the ways it has taken advantage of its riverine resources. She names the United States among those superior nations that have succeeded in this regard, since "the history of [a] country, economic and political, is indissolubly connected with that of its great rivers" (*Influences* 342). Her 1903 *American History and Its Geographic Conditions* sets aside an entire chapter to discuss the role that rivers played in North America for the purposes of early settlement and expansion, discussing their general importance for transportation, commerce, craftsmanship, and urban development in multiple places. The Mississippi alone appears almost one hundred times in the text.

Hughes's poetics transforms this valence of the Mississippi. Semple highlights four rivers as important to America's development: the Columbia, Missouri, Snake, and Colorado. Only one appears in Hughes's poem: the Mississippi. But its significance is not tied to development of the American nation in the senses that Semple and others concern themselves with. Instead, the Mississippi serves as an emblem of the promise of racial equality brought into being by Abraham Lincoln's realization when traveling down the Mississippi that slavery needed to end, and it retains its linkages to the other African diasporic rivers through the anaphoric "I" of the poem. In the heart of America's central artery of development lies instead, paratactically, the entire African diaspora. Past readings of the poem have viewed the relationship among the four rivers as a historical trajectory or narrative of development from birth to slavery to emancipation that culminates in the promise, or failed promise, of Lincoln's Mississippi. But by looking at the spatial dimensions of the poem, we can also notice the ways that this configuration of rivers reframes a master narrative of America's history of development as largely external to itself and as anchored to Lincoln's landmark turning point as its lodestone.

Here, too, is where the implicit comparativist logic of environmental determinism becomes transformed by the experimental modernist project into parataxis and by an African diasporic sensibility. Semple generates a host of global and historical comparisons in her 1903 *American History* (for instance, the topographical mapping of such maritime races as the ancient Phoenicians and those living in America's New England coastal cities [120]); in one set of comparisons Semple draws out the significance of America's river systems by comparing major rivers like the Mississippi to both the Nile and the Euphrates, not surprisingly with an eye toward extolling the virtues of how America's rivers are placed in the "most desirable part of the temperate zone" (20). "The Negro Speaks of Rivers" generates exactly these sorts of comparisons, although unlike a text like Semple's, which ignores the Congo as a possible comparison point, Hughes's poem includes it with the calculated effect of replacing primitivist ideas of savagery with bucolic images of peace after self-sufficient work. At the same time, African diasporic connection might be able to replace the infrastructure of American national development and even supplant it.

Understanding Hughes's engagements with geographical knowledge, whether in academic or popular forms, shifts our understanding of the terms on which he grappled with primitivism itself. David Chinitz has proposed that, unlike white modernist writers like Stein or Eliot or European cubist painters like Picasso and Braque who were exposed in the earliest decades of the twentieth century to primitivist ideas through anthropological and ethnological studies, Hughes's primary form of engagement with primitivism was through its popularization in the 1920s. But Hughes's repeated exposure to environmental determinist ideas and the racial associations they supported constitutes a longer trajectory along which we might track his responses to primitivist constructions. It also provides us with a new route for plotting coterminalities between African American writers and white modernist writers like Stein and H.D. who participated in the same set of larger cultural conversations about the relationship between the landscape and human society. In the case of Hughes, the stance toward environmental determinism is more overtly antagonistic because of its fundamental racial overdeterminations, whereas for Stein and H.D., it is more ambivalent. Yet with all three writers, forms of geographical influence and comparison undermine the project of nationalism.

4

(Trans) Nation, Geography, and Genius

Gertrude Stein's *Geographical History of America*

It is possible to read Langston Hughes's "The Negro Speaks of Rivers" as engaging with but ultimately unsettling the epistemological power of academic geography. The 1936 *The Geographical History of America; or, The Relation of Human Nature to the Human Mind* demands to be read this way, although Stein's relationship in this text to questions of American identity are motivated by a different set of concerns about the American writer living in permanent expatriation and about the nature of writerly genius. Her attitude toward American exceptionalist narratives is more triumphalist, although any support of Americanness in *Geographical History* is undercut not because of the driving force of an oppositional political agenda—although there is a little of that—but because of Stein's larger metaphysical considerations as well as her linguistic experiments. Unlike Hughes whose anaphoric parataxis juxtaposes units of conventional, syntactically regular sentences, Stein's style in her *Geographical History* suppresses the axis of selection (word choice) and heightens the axis of combination (syntax) that typically balance one another in the formation of grammatically correct sentences.[1] Parataxis here drills down to the level of individual grammatical units. These experiments engineer and sustain at their core a kind of ineffable, transnational fluidity.

The purposeful, thoroughgoing engagements with the disciplinary practices of academic geography in *The Geographical History of America* are announced by the book's title. Its contents, however, both do and do not actually deliver a geographical history of America in the conventional sense, in part because of the multi-generic nature of the text. As the seeming disjunction between the title, *The Geographical History of America,* and the subtitle, *The Relation of Human Nature to the Human Mind,* already indicates, the text is as much a curious, provocative hybrid of genres as almost any of Stein's other works. It generates material observations about the American landscape and its social and political formations for which the text depends upon academic geography, but it is also a philosophical treatise, a prose poem, an autobiography, a detective story, an identity play, a puppet play, an investigation of the nature of poetry, a playful plug for Stein's genius, a contemplation about government, and a "private" dialogue that Stein was carrying out with Thornton Wilder (she addresses him in *Geographical History,* one assumes, with a playful Steinian wink, and he supplied the text with an introductory essay when it was first published). On the first page alone, the text begins with a seemingly neutral declarative statement that asserts Stein's auspicious seat at the table of American leadership: "In the month of February were born Washington Lincoln and I" (*GHA* 45).[2] This reflects upon the space-clearing benefits of death, launches a few of the distinctions the text comes to explore that separate the qualities of the human mind and those of human nature, and offers the first of many statements that reflect upon the nature of America: "In the United States there is more space where nobody is than where anybody is" (*GHA* 45). The text revisits these issues, and more, in a dizzying nonsequential and nonlinear fashion.

Stein composed her *Geographical History* between June and September 1935 on the heels of her very successful American lecture tour in 1934–35, which was spurred by the wild commercial success of her 1933 *Autobiography of Alice B. Toklas* (Dydo 588).[3] With Hughes, travel becomes a starting point for contemplating the nature of geography. The impressions that Stein gathered from this tour from physically flying above the American landscape fuel *Geographical History*; a common refrain that appears in this book derives from her experiences looking down from an airplane as she flew across the flat expanse of the Midwest.

Repeatedly throughout the text, Stein returns to aspects of America's geographical history, chiefly the central notion that the flatness of much of the American landscape has been crucial in shaping fundamentally wandering, liberatory aspects of American culture. This equation richly echoes the logic of landscape-culture influence embedded in environmental determinist rhetorics, in ways that have led Jessica Berman to explore Stein's *Geographical History of America* in relation to the work of Ellen Churchill Semple. Although it is unclear whether Stein was directly familiar with Semple's work, there are several kinds of striking parallels between the two figures that Berman observes: both were female writers and intellectuals, and just as Stein studied with luminary philosopher William James at the Harvard Annex as a nonmatriculated student, Semple studied with Friedrich Ratzel in Leipzig but was consigned due to gender to listening to his lectures from the next room (*Modernist Fiction* 230n34). The similarity between Stein's title and Semple's well-known *American History and Its Geographic Conditions* gives Berman further cause to explore these two texts together, although I find the most compelling evidence prompting speculation about Stein's direct knowledge of Semple to be the fact that Semple's revised 1933 edition of *American History and Its Geographic Conditions* (*AHGC*, Rev. ed.) appeared shortly before Stein's American lecture tour of 1934 and before Stein set to work on her *Geographical History* in the summer of 1935.[4] For Berman, Stein both borrows from and supplants the disciplinary field of academic geography in order to replace its linear teleological qualities with her own nonteleological textual wandering, a process through which Stein forges a new sort of cosmopolitan community. But many examples of the geographical history genre are nationalist projects, with Semple's *American History* one of the clearest illustrations of this orientation. This fact opens up an interesting set of questions about the particular ways that Stein's *Geographical History* navigates Americanness and about how writing not just from outside America but in France in particular impacts these negotiations.

The scholarly conversation about Stein has only recently begun to move away from approaching her work almost entirely through the lens of avant-garde modernism or of experimental, postmodern poetics, often with concerted attention to questions of authority, gender, and sexual identity.[5] Priscilla Wald was among the first to note this critical tendency to read Stein in relation to aesthetics to the exclusion of cultural studies concerns, particularly within an American Studies context, as she charts Stein's unremitting efforts in *The Making of Americans* to tell the story of America's struggle to accommodate immigrants within familiar narratives (*Constituting Americans* 12). It is surprising, though, that a scholarly field increasingly attentive to Stein's relationship to Americanness has mainly focused on this text and the *Autobiography of Alice B. Toklas*.[6] The little work that has been done on

Geographical History does illuminate Stein's relationship not just to personal identity but to national identity, as it recognizes the text's obvious consideration of Americanness. But it tends to assume that whatever concerns Stein might have had about balancing her expatriate status against an American identity were shored up by the success of the *Autobiography of Alice B. Toklas* and of her American lecture tour.[7] From this perspective, Stein's *Geographical History* performs American exceptionalism by celebrating her native land's unique forms of open wandering, by linking these forms to the ability of the human mind to escape history, temporality, and fixity and thus the ability to generate "master-pieces," and by positioning America as superior to France.

These assumptions of Stein's American fervor follow the contours of some of her own gestures as well as her friend and literary confidant Thornton Wilder's own estimations of *Geographical History*. Stein performed her Americanness through proclamations, for instance, of exclusively smoking American cigarettes (Wagner-Martin 139). She imbued her lecture tour with a spirit of triumphant homecoming, publically affirming her sense of herself as a writer and her awareness of having a distinctly American audience. Upon her return to Paris after the tour, Stein declared to an American Associated Press reporter in Paris, "Yes I am married. I mean I am married to America, it is so beautiful. I am going back to America sometime, someday not too long. I am already homesick for America" (qtd. in English 258). "America," as scholars have viewed it, functions here as a dodge for addressing questions of sexuality and gender and plays up the indeterminate nature of Stein's relationship to Alice B. Toklas in public perception (Goble 128). Such readings take the "American" part of the equation as a straightforward, unproblematic articulation of what Hugh English calls Stein's "deep figurative associations" to "both the place and the idea of America" (258). It is no surprise that Stein's often-quoted declaration that "America is my country, but Paris is my hometown" has often been taken to mean that her national identity just so happens to "happen" in Paris. Wilder, who met with Stein and corresponded with her about her *Geographical History* regularly as she worked on the manuscript, saw the text largely through this frame. Upon reading the completed manuscript, he communicated in his October 7, 1935, letter to Stein his enthusiastic praise: "Gertrude, Alice, what a grand book. What an airplane ride, what a quilting-party, what a spelling-bee." Part of his admiration responds to the antifoundational qualities of the book: "how jubilantly naughty of Gertrude to sail in and smash half the accepted ideas of the world and to make such havoc so gaily. . . . And then to shake her head over the fact that the important literary thinking is being done by herself." Part of it registers his swelling sense of nostalgia for America: "My country tis of thee. I always knew I loved it, but I never knew I loved it like this. . . . I was born into the best country in the world. Gertrude told me so." Such incited nationalistic fervor makes him decide to return home from Vienna: "Something's happened to me. I'm crazy about America and I want to go home. I'm going to leave here in about three weeks. . . . Yes, I'm crazy about America. And you did that to me, too" (qtd. in Burns and Dydo 61–62). Stein's *Geographical History* certainly conveys its enthusiasm for America in multiple ways. But patriotic excitement does not necessarily make the text an uncritical meditation on American culture and its current state of affairs. Nor can we assume that Stein's relationship to America is easy or a given.

While it might be tempting to view Stein's interest in such ideas as a straightforward exercise of an Americanness that happens elsewhere, as some of her best friends and readers do, we can locate a more complicated set of dynamics in Stein's relationship to questions of belonging that stem from her uneasy ability to sustain affective personal attachments to the American landscape. Gerri Reaves views *Everybody's Autobiography* as registering Stein's substantial anxiety about returning during her lecture tour to a place she did not recognize and about whether she could reconcile the differences between her writing self and her newly famous self. I argue that Stein works at manufacturing her own Americanness in the text of *Geographical History* by shoring up her relationship to the nation's landscape. The text communicates an expression of American exceptionalism at the same time as it intervenes in contemporary American policies that threaten the nation's best promise for the genius artist, a dual process which allows Stein to restore the prospect of a better version of America to itself. *Geographical History* associates a longing for forms of American attachment that can be accomplished with flat looking with a longing for flat formless form, which only Stein, and not the American political leadership of Franklin D. Roosevelt in America, can provide. In worrying explicitly about Roosevelt's direction in the 1930s, the text generates a form of comparative nationalism, pitting Stein's against his. But the problems of identity—which Stein struggles with in numerous texts and worries could be too easily imposed on America—necessitate the very flat formless form that she struggles to write. The indeterminate, wandering nature of Stein's text undermines American exceptionalism and the environmental determinist principles that buttressed it. In this sense, *Geographical History* stages the limits of compatibility between American exceptionalism, on the one hand, and the experimental modernist project and/or the generation of a masterpiece, on the other.

These operations reflect in part the material conditions of Stein's cosmopolitan status as someone who calls America her country, but considers Paris her hometown. *Geographical History* was, after all, written about America but in France. The realities of the French context of Stein's writing, which critics have not discussed at any great length, open up new possibilities for understanding the text's complex positioning regarding geography and nation. The second part of my argument asserts that France plays an important role in Stein's engagement with environmentalist geographic ideas at the same time that it becomes an agent that unsettles their central assumptions about geographic ontologies. Stein's ability to manipulate geography resonates strongly with the dominant geographic paradigm of the modernist period in France: the idea of possibilism developed by geographer Paul Vidal de la Blache, France's best known geographer. While Stein may have been exposed to the possibilist ideas that American geographer Carl Sauer began to develop in the mid-1920s, Vidal's paradigm was a long-established one in France. It recognizes the human ability to modify and operate independently of environmental conditions or constraints and so posits more human agency than environmental determinism does. Vidalian geography also describes cities like Paris as exceptional spaces within the broader expanse of the country that facilitate cosmopolitan, extra-national forms of exchange. While the former aspect of Vidalian geography provides a more compatible way to link geography and genius, the latter facilitates the text's longing for distanced flat looking that is not dependent upon spatial proximity and that might well happen afar. In other words, because Paris is Stein's hometown in the Vidalian sense, she can write about America in the way that she does. But any sort of geographic attachments, of course, are subsumed into the larger unfinished, wandering project of the masterpiece.

Stein's relationship to America as a place and a concept was complicated despite her loyalty, appreciation, and sentiment. Although visiting America during her lecture tour brought Stein the numerous joys of a receptive public, an active schedule of readings and interviews, and countless meetings with prominent politicians, writers, and celebrities, her homecoming forced her to consider in new ways the nature of identity, time, and memory. It also forced her to contend with the contradiction of maintaining affective relationships to the landscape of one's national origin while maintaining expatriate status. In *Everybody's Autobiography*, in which Stein recounts her

American lecture tour and her return to France, she records feelings of dislocation and unease returning to places once familiar like California, where she had lived for a number of years as a child. California is both the same and different from her memories, a significant enough disjunction that Stein declares "we in our Ford car left for our California, this had been California of course but not our California the California we had come from" (*EA* 284). Stein repeatedly focuses on the discomfort that comes from things that have changed, things that are out of place, suggesting that she registers something more than nostalgic loss or misrecognition upon seeing the ways that locations and scenery have changed over time. When Stein and Toklas drive around the San Joaquin Valley, which Toklas had visited as a child, seeing the familiar poppies growing in California "which we had not seen growing wild since we had been in California" was a sight that Stein says "gave me a shock to see them there, it began to be funny and to make me uneasy" (284). This sudden appearance of the delayed familiar, or the familiar out of place, disrupts Stein's sense of "cognitive mapping" in which the human body attempts to orient itself in relation to its surroundings (Jameson 51), causing repeated scenes of unease and discomfort.

These problems become increasingly acute the closer Stein gets to her former hometown of Oakland, California. Although certainly these episodes reflect upon questions of memory and identity that Stein explores in *Geographical History*, they also center on the nature of Stein's affective relationship to the landscape and through it to a sense of communal and national belonging. When Stein explains that they left for "San Francisco and Oakland there I was to be where I had come from" (*EA* 288), her "to be" formulation throws into prominence the idea that "coming from" somewhere is dependent upon a particular "where." But almost immediately following this reference Stein highlights the unsettling hilliness of the San Francisco Bay area:

> all together the hills they had been and a great deal of them up and down we went among them and they made me feel funny, yes they were like that that is what they were and they did trouble me they made me very uncomfortable I do not know why but they did, it all made me uncomfortable it just did. (*EA* 288)

More than just the physical fluttering of driving up and down the steep hills that "make [her] feel funny," this experience made her "very uncomfortable" in ways that she both feels the need to repeat and also takes pains to confess that she cannot explain. Stein's uncharacteristic inability or unwillingness to identify her experiences fully may stem from an inability to reconcile an uncanny connection between the topography of the Bay area and the hilliness of France, which *Everybody's Autobiography* stresses repeatedly. In the middle of sketching her and Toklas's attempts to find new household help and recounting visits from friends, Stein offers an unprompted reflection on the beauty of the French landscape, noting that "[t]he hills are high" (*EA* 63). France's vertiginous terrain also punctuates the walks that Stein took in Belignin, the French countryside where she tended to spend her summers, with Bob Davis, a philosopher located in Vienna, and Thornton Wilder, who was also installed there while she was working on *Geographical History*: "So we went up the hills and down again there are lots of hills around Bilignin and we talked about the relation of human nature to the human mind" (300–301). After worrying about the prospect of Wilder serving as her literary executor, Stein decides, "I am not leading him I am confiding in him and that is what we did in going up and down the hills near Bilignin" (301). As I will discuss below, these aspects of topographical similitude start to bubble through hard and fast national distinctions between America and France that the *Geographical History* begins to question.

These feelings of dislocation and vertigo peak when she arrives in her former hometown of Oakland:

> I asked to go with a reluctant feeling to see the Swett School where I went to school and Thirteenth Avenue and Twenty-fifth Street where we lived which I described in The Making of Americans. Ah Thirteenth Avenue was the same it was shabby and overgrown the houses were certainly some of those that had been and there were not bigger buildings and they were neglected, and lots of grass and bushes growing yes it might have been Thirteenth Avenue when I had been. Not of course the house, the house the big house and the big garden and the eucalyptus trees and the rose hedge naturally were not any longer existing, what was the use. (*EA* 291)

Stein experiences extreme discomfort with this set of perceptions that stem not only from the disjunctions in identity and memory that divide her past self from her present self, but also from a sense of spatial disjunction as well. What is "the use" of having lived in Oakland if it only "might have been" the same Thirteenth Avenue where she was as a child? We can also read this as a direct reflection of anxieties about connection to landscape and to national identity. She goes on to ask, "What was the use of my having come from

Oakland" when "it was not natural to have come from there," which leads to her often-quoted dismissal of that area: "there is no there there" (*EA* 289). While this phrase has often been seen as expressing ideas of negation—as in, there is not anything there—it also registers a failure to connect to Oakland as a distinct place and as a landscape to which she can feel a sense of affiliation. I realize that my discussion has focused attention on Stein's relationship to Oakland to the exclusion of other American spaces to which she had claimed a past relationship, but several factors elevate its significance: it is at the western edge of America and at the end of Stein's journey, and it is also the place of Toklas's birth. Moreover, the comprehensive discomfort of this final California episode is signaled by her straightforward declaration: "I did not like anything that was happening. Later much later all that went to make the Geographical History Of America that I wrote" (*EA* 291). Stein's *Geographical History* seems poised, then, to address questions of temporality, memory, and identity *and* of vexed national spatial attachment.

In order to confront such questions, Stein animates her personal, emotionally charged reflections on the American landscape through inventive, subversive play with academic and scientific discourse. The title, *The Geographical History of America*, signals an overtly disciplinary attempt to pit writing—in particular the writing of masterpieces—directly against the discipline of geography and the genre of the geographical history. But we can understand these efforts to engage with material geography while also liberating "geography" from some of its typical classificatory and academic functions as part of a longer trajectory in Stein's career. In "Geography," a short piece written in 1923 in Nice, Stein distinguishes between "geography," as a way of categorizing and understanding the landscape and as a discipline of knowledge, and "geographically," a way of experiencing or performing the landscape. The very beginning of the piece introduces "geography" in its more conventional material sense that refers to Nice, albeit in a Steinian pun: "Geography as nice. Comes next geography. Geography as nice comes next geography comes geography" (*Painted Lace* 467). Here Stein also captures the ways a traveler in a new environment is presented with more and more geography or is presented with the alternations between the geography one sees and the geography on a map or a guidebook's descriptions of geography, narrated into a sense of temporal order in which one "geography" comes after another: "As geography return to geography, return geography. Geography. Comes next. Geography. Comes. Comes geography" (467). This rhythmic repetition of the word "geography" also enacts the movement of the typewriter (inevitably, Alice B. Toklas's) as it moves down the page. Movement in this portrait is Stein's central preoccupation, concerned as she is with the movement of the ocean waves off the shore of Nice. In the middle of the portrait Stein uses conventional definitional meanings to convey the action of water, although in a transformed adverbial state that better captures the fluid action of water as Stein experiences it: "Waterfully when the water waterfully when the water comes to soften when the water comes and to soften when the water and to soften" (468). Stein then uses the opposite of water—land—to express from a different perspective this rhythmic play of the ocean's waves perpetually inundating the shore: "Geographically and inundated, geography and inundated, not inundated" (468). Throughout the portrait Stein seems to be trying to move beyond "geography" in its fixed sense, declaring its limitations: "Geography pleases me that is to say not easily. Beside it is decided. Geographically quickly. Not geographically but geography" (468). Although Stein seems to offer up "geographically quickly" for consideration as an option similarly displeasing, this is quickly negated: "Not geographically but geography" is what does not please her easily. She seems to anticipate her readers calling her to task for replacing conventional usages of "geography" with the adverbial action of "geographically." As the portrait suggests at its close, "geographically" represents the sorts of wandering and spontaneity consistent with immediate action and the sense of presence earlier celebrated in the portrait: "Geography includes inhabitants and vessels. / Plenty of planning. / Geographically not at all" (470). Stein's writing works to capture the experience of the present in the act of travel (the sound and appearance of the ocean, the colors of the surroundings, the action of a traveler in sending postcards: "post" and the "post mark," "saving stamps" [469]), in part by enacting the geographic environment as a geographic space within her writing itself. While geography conventionally understood includes "inhabitants and vessels" and "planning," Stein's portrait and the process of living "geographically" contain these things "not at all."

Overt forms of the deconstruction of geography as a disciplinary practice and a type of knowledge project emerge in several pieces in her 1928 collection *Useful Knowledge*. There, pieces such as "The Difference between Inhabitants of France and Inhabitants of the United States" and "Wherein Iowa Differs from Kansas and Indiana" play upon the expectation of geographical differentiation and its relationship to defining distinct cultural and social traits only to reveal those differences to be meaningless and arbitrarily drawn, with the names for regions merely arbitrary signifiers.[8] Both Stein's lecture tour in America and her concern about government economic policies propel her to work out of the vein of academic geography in a more deliberate, sustained way in the *Geographical History*, which gives her a foundation for a connection that she exploits in at least a partial sense. Furthermore, though Stein's earlier geographically inflected works exhibit a nascent suspicion about the links between landscape and social and cultural traits, she explores the possible extremes of identity and creation—as they intersect with nation and landscape—most explicitly in *Geographical History*.

Many of *Geographical History*'s interventions regarding geography hinge upon Stein's metaphysical contemplations of the relationship between the human mind and human nature, which is a central formula that informs her approach to the American landscape, ideas of American government, the nature of the literary masterpiece, and ideas about belonging. As scholars have noted, the human mind and human nature are not entirely or neatly counter to each other throughout the text, although Stein does regularly position them in opposition and explicitly tips the scale to privilege the human mind.[9] Human nature is essentially being-in-time, characterizing daily life as it is lived. It is rooted in time through its association with memory and history, and it is associated with identity because identity depends upon forms of external validation that are similarly embedded in time and space. Stein's often-used formulation for identity, "I am I because my little dog knows me," appears in *Geographical History* in multiple forms to demonstrate how identity is fixed and static, molded by its relationality to other perceiving beings that recognize identity as such, by other social bodies, or by other externally produced forms of recognition. Identity is also constrained by temporal progression or by cause and effect, registered by such dismissive questions as "What is the use of being a little boy if you are growing up to be a man" (*GHA* 50). Human nature, or the identity it suggests, is rendered a limited category by its subjection to change and evolution. The human mind, on the other hand, transcends identity, history, and temporality: "there is no time and no identity in the human mind" (175). Stein claims for the human mind the ability to function free of time in part because it "believes in a glance and also in looking" (163). These particular forms of seeing involve a kind of "flat looking" that is not mired in history, a detachment that is central, in Stein's mind so to speak, to pure creation. "Seeing everything as flat. / When you look at anything and you do not see it all in one plane, you do not see it with the human mind but anybody can know that. It is naturally that. And so it is because there is no time and no identity in the human mind" (175). Playing on "plane" as a flat surface and also the airplane from which Stein looked out over the middle of America during her lecture tour, this type of flat looking can see the entire whole: "seeing everything as flat," seeing "it all in one plane."

Genius and its ability to produce masterpieces are linked to this sense of the human mind, operating at the height of its powers in a way that is free of the constraints that limit human nature: "The human mind has neither identity nor time and when it sees anything has to look flat. That is what makes master-pieces makes a master-piece what it is. And when it is only that only no time or identity then it is that" (*GHA* 175–76). Stein's own *Geographical History* attempts to fulfill this notion of a masterpiece on several levels. The constant recursivity and repetition-with-a-difference of Stein's phrases and formulations work against the temporality and causal logic of the conventional sentence, as does the text's disruption of the conventional mathematical sequence of chapters. To take just a few examples of this latter tendency: various chapter designations appear out of chronological order and are used repeatedly and seemingly at random (for instance, "chapter one" begins on pages 49, 51, 60, 132, 188); they cancel each other out (one of the early chapter IIIs that appears in the text, Stein declares, "is the same as chapter XV" [52]); and they mix roman, arabic, and spelled-out numbers, which creates, in Elliott Vanskike's words, "an arithmetic babel" (160). Stein's writing thus generates its own form of "flat looking," as it flattens temporal order and history. It enacts, as she articulates near the end of *Geographical History*, the idea of content without form: "What is necessary now is not form but content. / That is why in this epoch a woman does the literary thinking" (215). It is in this form of textual wandering that results that the text attempts to operate as a "master-piece": something that is the product of genius but that is also a "piece" that "masters" being and language in new ways. As I will discuss, this also enacts a distinctly American form of writing in step with the first part of Stein's often-quoted notion that "America is my country" but "Paris is my hometown" ("An American in France").

Stein links these transcendent, liberatory qualities of the human mind to the flatness and largeness of the American landscape, a move that depends on environmental determinist principles. Ideas of westward expansion were the central means of constructing American identity in the nineteenth century, captured most visibly in Frederick Jackson Turner's frontier hypothesis, which Semple's 1933 *American History and Its Geographic Conditions* makes clear was a popular understanding as early as 1841 (220).[10] At various turns,

Semple's book creates almost a reverse sense of manifest destiny such that the American landscape itself facilitated national expansionism rather than territorial expansionism functioning as an expression of the nation's will. One of the key features to which Semple attributes America's remarkable growth and vitality is the vastness of the landscape:

> Mere space, unconstrained existence, a buffalo hunt, or an Indian fray was pleasure enough. In the large, fresh environment of the American continent the English race had been born again and now was animated with the irrepressible vigor of a youthful people. A constant change of environment had given them the adaptability of youth, vast opportunity had bred the spirit of venture and enterprise. Nothing seemed impossible and therefore little was impossible. (*AHGC*, Rev. ed. 231–32)

This sense of unchecked space as a central constituent of Americans as a new and adventurous society informs a similar linkage in *Geographical History*:

> Now the relation of human nature to the human mind is this.
> Human nature cannot know this.
> But the human mind can. It can know this.
> In the United States there is more space where nobody is than where anybody is.
> This is what makes America what it is.
> Does it make human nature in America what it is. If not it does make the human mind in America what it is. (45–46)

The precise definition of the relation between the landscape and the human mind shifts and changes throughout the text, sometimes portrayed as a causal relationship, at other moments as an open-ended sort of association in which a vague "something" connects them: "The land has something to do with the human mind but nothing to do with human nature" (*GHA* 109). But Stein is clear in asserting that the human mind can know America's spaciousness and emptiness in ways that human nature cannot. This is presumably because the human mind is made by America's flat, empty space, a point of clarification Stein makes in ways that echo the moves of the strictest form of environmental determinism that would claim that the environment "makes" human culture: "Does it [the riches of America's empty space] make human nature in America what it is. If not it does make the human mind in America what it is." Although here Stein brackets this claim within the conditional parameters of a "what if" statement, and elsewhere in *Geographical History* notes that not all inhabitants of a big, flat country wander, she repeatedly articulates here and elsewhere the idea that the human mind is fostered by, or even produced by, this particular set of environmental conditions.

The flatness of the American landscape, made visible by the genre of the geographical history, is so central for Stein because it promotes the sort of wandering necessary for the creative operations of the human mind, as opposed to adhering to fixed linear, teleological trajectories:

> Wandering around a country has something to do with the geographical history of that country and the way one piece of it is not separated from any other one. Can one say too often just as loving or tears in one's eyes that the straight lines on the map of the United States of America make wandering a mission and an everything and can it only be a big country that can be like that or even a little one. (*GHA* 84–85)

Stein's language of "wandering" echoes Semple's attention to the "migratory instinct" of Americans in the frontier that geographical conditions "favored" (*AHGC*, Rev. ed. 229). This set of conditions encouraged settlers to "take up unexhausted lands, an unused 'range,' and also to seek that unconstrained life of the backwoods which has a powerful charm to the natural man" (*AHGC*, Rev. ed. 229), allowing them the freedom to explore and also freedom from the limits of daily life in smaller, more developed urban locations. Stein figures this lack of environmental and social constraint through an image of indivisibility and interconnectedness, "the way one piece of it is not separated from any other one." This, too, finds its parallel in Semple, who asserts that in the "natural" advance of America from one coast to the other "the large and simple structure of the continent kept the Americans one people" (*AHGC*, Rev. ed. 232).

In framing these uniquely American qualities in this way, Stein affirms the American exceptionalist qualities in Semple. Although Stein's celebrations of "wandering" seem to work against the linear nature of America's national push westward, the total effect, the distant "flat view," rests with the exceptional social, cultural, and geopolitical benefits that these wide open flat spaces create. Stein channels the sense of American global superiority that Semple's work repeatedly articulates. Although Jessica Berman's discussion of Semple stresses the regional qualities of *American History and Its Geographic Conditions,* these particularities are subsumed by Semple's larger efforts to define and celebrate America not only as a nation but as a nation with a large role to play on the global stage. *American History* makes frequent assertions regarding the geographical benefits that have allowed America to establish its strong international presence; for example, because of its spaciousness and its temperateness, the "central belt of the North American continent was geographically determined as the seat of control of the new hemisphere." This destiny was fixed by its colonization by the sturdy Anglo-Saxon race" (*AHGC*, Rev. ed. 239–40).[11]

Stein builds on this sense of superiority directly, while in more subtle ways employing the comparativist logic of environmental determinism that relies upon spatial differentiations and spatial comparisons in order to generate its claims about the cause-and-effect relations between environmental conditions and particular social outcomes. *Geographical History* makes several comparisons between America and European countries in terms of geographical and social freedom. In several instances, Stein not only asserts America's wandering strongly in *Geographical History* and enacts it in the text itself through repeated spreading and purposeful repetitions; she also says that France starkly lacks this quality: "The relation of nervousness to excitement and the death and the death of René Crevel. René Crevel was not nervous he really was not excited and that is because he was in a country where no one wanders" (85). Stein implicitly refers to the chronic health problems plaguing well-known French surrealist novelist René Crevel, which ultimately led to his suicide in June 1935 as Stein prepared to work on her *Geographical History*. Here, Stein attributes his problems and perhaps even his suicide to a lack of the freedom, interaction, and excitement of wandering. The danger to the artist of being in a country in which "no one wanders" constructs a clear global hierarchy that privileges America over its European competitors.

Although Stein celebrates these seemingly essential, privileged qualities of Americanness, this does not necessarily mean that *Geographical History* is purely laudatory. In fact, it tracks a widening gulf between these aspects of nation/American and contemporary governmental forms, particularly in terms of the reigning economic policies of the New Deal. The accretive chain of associations among the human mind, flat land, and the wandering qualities of America contrast with where these economic policies might lead, which Stein positions as distinctly out of step with American identity. Although the economic aspects of *Geographical History* have escaped scholarly attention, Stein presents the project of the geographical history itself as one that is central to understanding the nature of government and politics. On the very first page of *Geographical History*, she addresses "propaganda and politics and religion" and declares "A Geographical history is very important when connected with all this" (47). In this regard, Stein's *Geographical History* puts the monetary policies of Franklin Delano Roosevelt's New Deal under sustained scrutiny, suggesting that it can run counter to the sorts of free wandering that both the American landscape and Stein's writing embody. Stein's own particular take on the idea of a geographical history can liberate Americans from these bad policies and the intrusive, oppressive form of politics they constitute, restoring in a sense America's flat, wandering qualities of landscape to itself.

To dissect FDR in this way seems surprising, given the general terms of Stein's patriotism performed elsewhere in the *Geographical History* and her personal connections to the Roosevelts. During her American lecture tour, Stein took tea at the White House on December 30, 1934. Although Mrs. Ann Archibald, the organizer of Stein's lecture at the Women's University Club in D.C., had initially offered Eleanor Roosevelt an invitation to tea, Mrs. Roosevelt couldn't come but invited Stein and Toklas to White House for tea instead (letter dated 20 December 1934 from Lamotte to Stein, Stein and Van Vechten 371). Later that spring, Stein had dinner with Franklin D. Roosevelt Jr. in Fort Worth (Burns and Dydo 349). These personal connections may well shape some of Stein's central philosophical considerations of identity offered in the *Geographical History* as well as in some of her other texts, normally couched in terms of Stein's own identity: "I am I because my little dog knows me." In *Geographical History*, these considerations of identity incorporate Roosevelt's identity into the equation: "Well any Franklin Roosevelt has he any identity. / I am I because my little dog knows me. / But does any little dog know more than know that it is he" (138), and a few lines later, "Does the little dog know that he is he / But is he" (139). Drawing Roosevelt into these philosophical disquisitions would have tickled Thornton Wilder, who like other friends of Stein was a staunch Roosevelt supporter.[12]

But her economic politics combined the conservative with the provocative. Edward Burns and Ulla Dydo have discussed her economic conservatism. The fact that her long-standing friend William Rogers attributed to her the perspective of a "'rentier,' a person of property," shapes their view that she "opposed Roosevelt and the New Deal and was more afraid of communism than of fascism" (414). They also cite Stein's 1934 interview for the *New York Times Magazine* in which she claims that Hitler should receive the peace prize, the biting irony of which is only made clear later in the interview when she expresses support for the very things that Hitler's regime would suppress: "competition, struggle, interest, activity that keep the people alive and excited" (qtd. in Burns and Dydo 414). Regardless of her intentions regarding how Hitler should be considered, she ties these subsequent comments about the conditions that "keep the people alive and excited" to the very nature of the capitalist endeavor.

At multiple junctures in *Geographical History,* Stein considers Roosevelt's New Deal actions as an effort to get rid of money and wonders if this is a creative project that exists in its own right or a calculated set of "electioneering" gestures mired in history. She shows that this effort aims to produce a particular future outcome in a particular audience, both aspects associated with time and the fixity of identity that she aligns with human nature. It is important to note that she presents the notion of Franklin Delano Roosevelt's wrong-headed elimination of money not as a certainty or inevitability in the text but rather as a possibility. Even near the end of *Geographical History,* Stein allows the question of whether Roosevelt is getting rid of money to go unanswered: "I tell you it is true that I do the literary thinking for you. / Even perhaps if Franklin Roosevelt wants to get rid of money" (221). Such conditional "if" clauses about Roosevelt are typical throughout the text, although their repetition and subtle modulation take on a sort of accretive urgency about this as a threat. To suggest through a flat wandering pattern rather than assert a set of claims outright seems, after all, more consistent with the human mind's desire to look flatly at the bigger whole while avoiding fixity. "Is Franklin Roosevelt trying to get rid of money. That would be interesting, but I am afraid it is only human nature, that is electioneering and that is not interesting" (217). Whatever "interesting" qualities associated with eliminating money as a thing in itself as opposed to eliminating it in an instrumental way for political gain stem from the fundamental associations Stein draws between money and the human mind: "There is flat land and weather and money for the human mind" (*GHA* 101). Against this chain of associations Stein pits government and propaganda: "All this [identity] has nothing to do with the human mind but so much to do with history and propaganda and government but nothing to do with money and the human mind nothing to do with money and the human mind" (*GHA* 135). Electioneering and its perhaps inevitable tie to governmental work are linked to identity and human nature.

Regardless of Roosevelt's intention, his efforts to get rid of money disrupt its central role in contributing to the wandering nature of America's flat landscape. Although Stein never definitively identifies what she means by money, she does repeatedly assert that it is "interesting" (*GHA* 181) and characterizes it as mobile, free of time, and free of identity in ways that connect it to the human mind ("Really money really has to do with the human mind" [161]) rather than with human nature ("Money has no time and no identity and no human nature" [195]). Stein does not fetishize money per se but rather views it as central to pure creation. It operates in Stein's view much like the human mind because it is a system that can point to anything, and it does not depend for its meaning on the past or the future: "Money is the difference and money who likes money money is what we all agree, to be happy and make money, is anything" (182).

For example, Stein associates money directly with "words": "Money is what words are / Words are what money is. / Is money what words are / Are words what money is" (*GHA* 193). Stein may be winking at the reader with a Rumpelstiltskin-like set of syllogisms, given the fact that her wildly popular *Autobiography of Alice B. Toklas* published in 1933 had earned Stein quite a lot of money: her words had spun gold. Something of this sentiment playfully animates her claim that "I am writing all this with an American dollar pen" (183), an image that seems to blend together in one gesture the low-culture, crass commercialism of novelty patriotic items and Stein's efforts to write literary masterpieces in a way that America has fostered. One gets the sense that Stein's economic cushion from *The Autobiography of Alice B. Toklas* enables her to write a more experimental work like *Geographical History*. Most important, the equation between money and words also refers to the way that money—the word itself—multiplies and spreads across Stein's patterns of repetition and recursivity within *Geographical History*. But even considered apart from its linguistic performances, money as a system of exchange operates free of identity, history, or temporality, perhaps because it can, like language, represent virtually anything, in this case anything commodifiable under capitalism, as opposed to a barter economy where things as themselves are brought into a relationship only with the things for which they are traded. In this respect, money is free of the fixative effects of time and identity: "Money and words . . . have no time or identity in them, no certainly not" (194).

Money therefore has a central role to play in the chain of associations Stein creates that link America's flat land, wandering, money, and the human mind with the absence of government and human nature:

> Flat land is not romantic because you can wander over it and if you can wander over it then there is money and if there is money then there is the human mind and if there is the human mind there is neither romance nor human nature nor governments nor propaganda.
>
> There should be none of these if the land is flat.
>
> Flat land as seen from above.
>
> Above what.
>
> Above the flat land. (*GHA* 155)

If we read "There should be none of these if the land is flat" to refer to romance, human nature, governments, and propaganda, then Stein is suggesting that governments and propaganda—New Deal economic policies and the electioneering in which they participate—are incompatible with America. Recognizing this incompatibility is contingent upon the idea of flat land "as seen from above," implying that Roosevelt is unable to see the spreading unity of the whole and to see flatly in the way that the human mind and Stein can.

As a result of his inability to see, Roosevelt threatens to disrupt this important chain of wandering by attempting to get rid of money. Roosevelt's New Deal, in other words, is antithetical to the human mind and therefore antithetical to masterpieces. Such policies also threaten, by extension, the imposition of government, of top-down linear authority, in these flat, wandering spaces. The intrusiveness of this policy is registered by Stein as a potential affront to the free, wandering aspects of America's flatness. She clarifies a few sentences later: "And government, no there is no government where the land is flat. / There should not be. / And there is not. / Because anybody can wander and if anybody can wander then there should not be any human nature" (*GHA* 156). This threat does not seem to be a realized one: government should not be there and "is not" precisely "because" anybody can wander. While Stein generates ambiguity over whether government poses an imminent threat to the wandering qualities of flat land ("There should not be" government impinging upon flat land) or whether flat land always already evades the incursive machinery of government ("And there is not" government), the earlier placement of money in the chain of conditions suggests that New Deal economic policies could tip the scale in the wrong direction.

Stein diffuses throughout the *Geographical History* the concern about governmental policy that she expresses more pointedly in her portrait "A Political Series," also written in 1935, in which she pursues the nature of Roosevelt and his New Deal economic policies in more focused and compact form. There, she claims that Franklin Delano and Theodore Roosevelt are not American because "every nation has a way of being that nation that makes it that nation and anybody that is really important in the development of that nation has to be some way somehow like that. If they are not like that like their nation and they get to be important in the nation they lead the nation to a real catastrophe." Being American "has nothing to do with where they were born or where they lived" (*Painted Lace* 72–73). Stein expresses her sense of their disconnection specifically by showing their incompatibility with a country identified with wandering that exists below or apart from governmental structures: "I mean they do not feel America to be a very large country around which anybody can wander and so although a government is there it is not always anywhere near but they feel it to be a little country which they can govern, and so it is European and not American" (74).

We see Roosevelt's problematic disconnect from the bigness and wandering qualities of America in *Geographical History* as well, although packaged as evidence of or as related to his lack of a human mind. Franklin Delano Roosevelt, Theodore Roosevelt, Napoleon, and Louis Napoleon lack an ability to "see" and "use" land properly because they lack the human mind: "They saw the land they could use but they could not use land and as they could not use land they could see land but as they saw land what land well not any land because after all land is land, that is the human mind and they had no human mind" (*GHA* 130). Lacking the ability to "see" and to relate to the land makes all of these figures outcasts from their own country's landscape: "Theodore and Franklin Roosevelt like Napoleon and Louis Napoleon even though they belonged to the country to which they belonged were foreign to it" (127). Here Stein implies a distinction between official citizenship and a more meaningful form of connection and belonging that is not dependent upon actually being physically present and that operates purely through flat looking.

FDR's disconnection from the nature of the American landscape and all that it fosters sets up the prospect of an intrusive governmental style that would interfere with America's essential qualities. In *Everybody's Autobiography* Stein expresses Roosevelt's impact as a generally oppressive form of fathering, of which America has too much:

> There is too much fathering going on just now and there is no doubt about it fathers are depressing. Everybody nowadays is a father, there

> is father Mussolini and father Hitler and father Roosevelt and father Stalin and father Lewis and father Blum and father Franco is just commencing now and there are ever so many more ready to be one. Fathers are depressing. England is the only county now that has not got one and so they are more cheerful there than anywhere. It is a long time now that they have not had any fathering and so their cheerfulness is increasing. (133)[13]

Geographical History suggests that this form of governmental intrusion threatens to limit the expression of the human mind. Government in general is loosely associated with the ability of the human mind to say yes, which defines aspects of the human mind's function: "it is very exciting but think how much America and I do think America has something to do with the human mind think how much America has to do with yes" (123). While it would seem that Stein is affirming government styles and its related actions by associating it with the "saying yes" of the human mind, a notion that would negate the suspicions about the impact of Roosevelt's New Deal economics (which Stein might place, problematically or not, on the slippery slope to communism) for the sake of electioneering (propaganda), the implication is either that these political formations and practices need to be generated from the "yes" of human minds asserting themselves from below, rather than imposed from above, or that once these systems are imposed, "all the human mind can do is to say yes." Regardless of the particular meaning, Stein makes it clear that America is suffering from a "yes-saying" deficit and is risking becoming mired in human nature as a result: "Poor America is it not saying yes, is it loosing [*sic*] the human mind to become human nature. Oh yeah" (125). By affirming this with a "yeah" rather than a "yes," Stein playfully enacts the enervation of these "yes-saying" energies.

Stein's *Geographical History* helps to restore this threatened connection between America's own flat land and the human mind by performing these connections with her own flat, wandering writing. Unlike Roosevelt, Stein is one of those people who, from "A Political Series," is "really important in the development of that nation [who] has to be some way somehow like [the nation]" (*Painted Lace* 72). She is able to make her text the delivery vehicle for her own brand of geographical history that can restore America to itself. This formula is expressed in terms of *shoulds* and *oughts*, to return to the passage cited above:

> Flat land is not romantic because you can wander over it and if

> you can wander over it then there is money and if there is money then there is the human mind and if there is the human mind there is neither romance nor human nature nor governments nor propaganda.
>
> There should be none of these if the land is flat.
>
> Flat land as seen from above.
>
> Above what.
>
> Above the flat land. (*GHA* 155)

Stein presences all of these components in her text: the flat land of America is something that the recursive, repetitive nature of the text "wanders over" and sees "from above" but in ways that do not fix the landscape into geographic categories or into the realm of identity or time associated with human nature. *Geographical History* itself performs the formula that "[t]here should be none of these" (romance, human nature, governments, propaganda) if "the land is flat" and is flat "as seen from above" in the way that Stein's human mind allows her to do, or at least in the way that Stein hopes the human mind allows her to do.

By claiming and making available her own human mind to the reader, Stein creates a sense of community with her reader that is linked by shared access to the human mind, which she has associated with flat landscapes and hence America. Her very own human mind appears on the space of the page in a performative act that brings it into being so that the reader may come into contact with it:

> Please see my human mind.
>
> It is here.
>
> Is white a color.
>
> Yes white and grey is a color.
>
> Grey and white is a color. (*GHA* 120)

Stein instantiates her own human mind within her own text in a way that is both playful and meaningful. The "here" can quite literally refer to the writer's own head with its shock of salt and pepper hair, which in turn physically embodies on the space of the page Stein's physiological human mind as the thing inside her skull. But it also merges this sense with the ink and white of the pages of her book. Her book becomes the human mind as a textual landscape that the reader can apprehend. If Roosevelt is disconnected from America's land, and Stein is both connected to the land and able to activate that land within her text, then her readers are able to access the human mind and the flatness of the landscape through her work. This logical chain returns us to the very beginning of *Geographical History*, where Stein introduces herself among the notable American leaders George Washington and Abraham Lincoln: "In the month of February were born Washington Lincoln and I" (45). The justification for this linkage initially seems dependent upon temporality and history because February is their shared birth month. But through the process of the text's negations of temporality and history, her associations with them take on a different cast that seems more a reflection of their common ability to connect to America.

Although what I have outlined thus far suggests that Stein asserts her relationship to the human mind and all that it is connected to with absolute certainty, this is something that the text approaches with a characteristic playful deconstructive tendency that both presents and takes away, asserts and negates. Some of her most direct assertions of her connection to the human mind are also the most playful. Anticipating readers who might take her to task for the proliferating characteristics she attaches to the human mind in the text, Stein addresses these frustrations directly, as she often does in her writing, by declaring her sanity while simultaneously using the literal sense of possessing a human mind to subtly argue, once again, her own connection to the human mind: "I am not confused in mind because I have a human mind" (*GHA* 153). Later in the text, certainty about the human mind (or her human mind, thinking), however humorously rendered, is undermined at least partly by a question about whether Stein has lost a grip on what the human mind actually is ("And all the time was I right when I said I was losing knowing what the human mind is" [*GHA* 222]), a grip that has existed "all this time," one can presume, across the temporal span of writing the entire text. Even the commands Stein issues about searching for the human mind, "Begin being ready to find the human mind" (*GHA* 122), are couched not as a certainty but as a conditional future prospect; the human mind appears as something for which readers can only prepare to be ready, a future possibility rather than immanence awaiting discovery.

These ideas of uncertainty also impact Stein's constructions of space and national identity, which allow her to deconstruct some of the basic reigning environmental determinist assumptions of the day directly and with them essentialist ideas of an exceptional American identity. Stein tests the limits of Semple's kind of formulation first by prying open the disjunctions in scale between the aggregate view of entire communities that environmental determinism takes and the micro-level scale of individuals. Stein points out that not everyone in a large country wanders and that the aggregate does not always scale downward completely: "In wandering around a big country some people who live in a big country do not wander" (*GHA* 85). More important, the categorizations of America as a distinct, bounded community themselves bend under Stein's treatment as she erodes the distinctions between America's capacious wandering and smaller countries like France that lack these qualities. Almost in direct conflict with other statements that confidently proclaim America's bigness as key to its wandering-ness, Stein explains, "But really wandering has something to do with the human mind. A big or little country. Wandering in a big or little country" (85). The linkage between the human mind and America as a big, flat country made elsewhere in the text is replaced by an "or" that offers up a "big or a little country" as equally plausible contexts in which wandering can occur, suggesting these possibilities exist elsewhere, even in France, which the text tends to present as a prototypical small country. Elsewhere in the same passage, she wonders more overtly, "can it only be a big country that can be like that or even a little one" (85), questioning whether wandering can in fact take place in a small country like France as well as in a large one like America. Thus the exceptional qualities of America do not remain uninterrogated in the text. Here Stein might be implicitly testing Semple's formulations against her own personal experiences in the French countryside as she was writing *Geographical History*, roaming the hills of Belignin as she talked with Thornton Wilder and the philosopher Bob Davis, both of whom were visiting from Vienna. Her recounting of this set of experiences in *Everybody's Autobiography* highlights two qualities of France that implicitly contrast with the flatness of the American landscape. The smallness of France strikes her anew upon her return, as "everything looked little and littler than it had looked" (*EA* 297), and the hilliness of the terrain in Belignin features prominently in the text. Recounting visits from Davis and Wilder, Stein notes their walking and talking: "So we went up the hills and down again there are lots of hills around Bilignin and we talked about the relation of human nature to the human mind" (*EA* 300–301). Most telling is that Stein articulates her sense that although she is "slow minded and quickly clear in expression," "I am clear I am a good American" who can "see everything that is seen and in between I stand around but I do not wait, no American can wait" (*EA* 301). *Everybody's Autobiography* thus performs the ways that Stein can wander, both literally and intellectually, as she goes on long walks with her interlocutors.

Alongside this effort to reverse or evacuate national comparisons, Stein explores an even more radical move in which the ontological fixity of material space is directly questioned and along with it America's singularity, its ability to produce or foster "more" human mind than other countries and its ability to generate masterpieces: "Is there any difference between flat land and an ocean a big country and a little one" (*GHA* 215). The topographical features that geographers would take as their starting point for their equations of environmental influence and social response can no longer function in such a world in which flat lands, oceans, big countries, and little countries cannot be told apart ("is there any difference" in the sense of identifying these things as distinct entities), or if they can in fact be differentiated that those differentiations ("is there any difference" in the sense of what difference it makes) are meaningless. Stein immediately follows this question with a similar one: "Is there any difference between human nature and the human mind." While this sort of probing threatens to undercut the viability of almost all of the associative chains the text has thus far laid out (the human mind, money, wandering, flatness, America's large open spaces), these questions of categorization and ontology become subsumed into the formless, wandering qualities of writing itself. Just a sentence later, Stein reminds us once again that "What is necessary now is not form but content" (*GHA* 215). Here Stein seems to be putting in play a sense of form as the forcing of categorizations and distinctions central to the act of putting things in a sequence or giving them a shape that unfolds temporally, with a specific kind of beginning, middle, and end. Her refusal to answer her own questions explores the notion of pure presentation of things but in a way that is free of fixing their "thingness." To offer the only answer to her questions about landscape and the human mind and human nature as an assertion that what is needed is "not form but content" seems of a piece with this sentiment, since questions and answers are themselves, for Stein, "form." If these things are true, then formless content replaces the sequential, temporal aspects of questions and answers with the pure presentation of possibilities.

Although such propositions would make it seem almost impossible for Stein to offer any kind of stable or meaningful "meaning" in *Geographical History*, these efforts to undo categories and to evade form, however, are consistent with the incomplete nature of the masterpiece as Stein defines it: "In writing [versus "saying"] not any one finishes anything. That is what makes a master-piece what it is that there is no finishing" (222). The notion of the masterpiece as unfinished or as unfinishable is also linked to money and romanticism, both of which "do not end and they do not begin" because "they do not exist and therefore they are not anything" (*GHA* 223). Whatever sort of geographical history Stein inherits, and the "flat looking" at America it allows her to generate, are subsumed into a wandering, processual, and unfinished text.

Stein's efforts to use but also undercut American exceptionalist ideas and the environmental determinist principles that support them is of a piece with her systematic efforts to "smash half the accepted ideas of the world and to make such havoc so gaily," as Thornton Wilder put it in his letter to Stein quoted above. But the geographical epistemologies with which Stein engages may not be only American in orientation. While it is unclear whether Stein read French academic geographic work, it seems hasty to assume that she was not familiar with the major trend to overtake geography in the late nineteenth century through the 1930s in the country in which she had been residing for decades: the "possibilism" of Paul Vidal de la Blache, frequently referred to as the father of modern geography in France (Berdoulay 52). Because of her wide-ranging tastes and far-reaching intellectual circles, it is quite likely that Stein draws upon the general contours of possibilism in connecting with America at the same time that she registers her experiences of "American-like" qualities in France. Articulated in his best-known work, the 1903 *Tableau de la Geographie de la France,* Vidal's possibilism advanced the idea that nature performs more of an advisory role rather than a deterministic or controlling one, which allows the human to play an active role in the shaping of society (Gregory 2). The man-land equation, in Vidal's view, operates more as a dialectical relationship than an antipodal one and positions the human aspect as a free agent who can operate as a "free user of the earth," to cite Vincent Berdoulay's characterization. In fact, Vidal and his followers stressed the human creative principle in particular that benefits from but can transform the natural environment. "Hard work, initiative, association, and social solidarity" were all celebrated as a part of this dialectical relationship, according to Berdoulay (58).

This possibilist focus on the creative use of geography aligns with Stein's privileging of writing in *Geographical History*. Her own project pursues "writing," which is an action, a sort of doing of things with geography. Part of this depends on the distinction Stein draws between "saying," which is temporal, embedded in history because it depends upon a particular audience, and related to identity, and "writing," which is free of temporality, history, and identity: "as I am never writing what I am saying when I am writing I am as it were not saying something and so then there it is that what writing is not saying something content without form but anyway in saying anything there is no content but there is the form of question and answer" (*GHA* 227). "Saying" is not content-ful but instead is nothing but form, to which the *Geographical History* seems antithetical; "writing" on the other hand is "not saying something" and is "content without form." The approach to geography that *Geographical History* seeks to reject is "saying" something about geography as the geographer would, which imposes categorization and stasis ("form") upon it. Instead, the pathway to geography that Stein pursues is one in which she "write[s] it so," which avoids those fixative tendencies:

> Why the writing of to-day has to do with the way any land can lay when it is then particularly flat land. That is what makes land connected with the human mind only flat land a great deal of flat land is connected with the human mind and so America is connected with the human mind, I can say so but what I do is to write it so. Think not the way the land looks but the way it lies that is now connected with the human mind. (*GHA* 79)

The way that land "lies" is not the land circumscribed by the geographer or by human nature generally but is the land that "lies" on Stein's page, that Stein brings into being in her flat, formless, and wandering text. This performative enactment of geography becomes something that Stein does to the land, even if that means turning material land into a phenomenological concept. Here Stein brings this connection into being in a new way and draws it into the phenomenological and literary act of writing itself, rather than repeating any of the principles "said" by geographers, whether they are Semple or even Vidal. She returns to the performative sense of geography that she explored in her piece about Nice, albeit here she is enabled by, and then surpasses, the possibilities of possibilism.

French geography also provides a broader intellectual context for understanding the ways that Stein might be able to write *Geographical History* about America, although she is in France, and why in the broader sense Stein might be able to navigate between two countries' schools of geography. Vidalian geography tended to accord cities like Paris, which after all is Stein's "hometown," a cosmopolitan character that bordered on being an exception to the French nation. This approach to Paris helped to make legible, if not to enable, Stein's ability to link with American culture from her outside position. Two Vidalians, Millicent Todd Bingham and Raoul Blanchard, articulate these ideas in the well-known 1919 *Geography of France*. The book's authors, like Stein and Toklas, straddled French and American contexts.

Bingham was, like Semple, one of those rare women geographers in the early decades of the twentieth century, having received a PhD in geography under special dispensation from Harvard in 1923; she also translated Vidal's *Principles of Human Geography* into English in 1926, giving that text new life and introducing it to English-speaking readers (Harvey and Ogilvie 127). Bingham and Blanchard draw attention to the way that France is divided up into a dozen or more distinct basins, and they explain that the Paris basin, unlike the more rural ones, is the "meeting point" of a variety of ideas and cultural flows. From this "rich interweaving" emerges "the national spirit, of which the high expression is the Frenchman's ability to keep in sympathetic touch with the ideals and thoughts of other nations" (8). If, as Bingham and Blanchard say, the regionalist geography that has emerged from Vidal onward focuses on the "battle *within a given region* between man and nature" (12) in which the human component has the capability of submitting nature to its will, they, like Vidal, position Paris as having evolved to a higher state of affiliation such that the focus of Parisian society is no longer the relationship to the landscape but the relationship to social formations external to it. [14] Paris, in short, has evolved past this battle and is able to generate a cosmopolitan, extroverted outlook. This perspective provides a way to envision Stein's Parisian sense of belonging as not incompatible with her connection to America. In this respect, *Geographical History* can be seen as positioned disciplinarily between the American geographical contexts upon which Stein draws and the contributions to geography generated by French sources.

5

H.D.'s *Trilogy* as Transnational Palimpsest

Stein's (anti)comparative nationalism grows in part out of her plane's-eye view, the perspective of a traveler who quite literally sees geography from a different and distant perspective or angle of vision. For H.D., who decided to stay in London during World War II and therefore was subjected to England's restrictions on travel, transnational transportation on this scale was not an option. Indeed, the opening image of *Trilogy* focuses on the transformations of local conditions under the sign of war: "rails gone, (for guns)" (*CP2* 509).[1] H.D.'s primary reference with "rails" is the British war office's melting of the wrought-iron fencing around her Knightsbridge neighborhood's squares and gardens for munitions, which opens up urban geography in new ways. Whatever physical, literal travel H.D. and her speaker could engage in is distinctly local. And yet books—about Egypt and the Mediterranean, in particular—became a form of textual travel that significantly shaped H.D.'s approach to *Trilogy*. Through H.D.'s reading habits we can observe a clearer line of sight to the work of Ellen Churchill Semple than we can observe with Stein, and yet both writers use forms of environmental determinism as a strategy for testing the limits of nation and, more so in H.D.'s case, of empire as well. This, notably, converges with a paratactic disruption of syntax and narrative that both writers use to test national boundaries and hierarchies.

Ellen Churchill Semple's *Geography of the Mediterranean Region: Its Relation to Ancient History,* published in 1931, was the labor of two decades of research and travel and the product of dogged determination that allowed her to complete the work despite major health problems (Creese 317). Drawing heavily upon ancient literary and philosophical texts to illuminate what the geography of the Mediterranean looked like and the larger economic, technological, and cultural dynamics it could reveal, the book's spatial and historical orientation allowed Semple a joyful return to her early academic roots in classical history. It failed to reach the same degree of prominence among academic geographers compared with her earlier works in part because the intellectual tide in the field had started to turn by the mid-1920s: at least within those academic circles, environmental determinism was by then an outmoded model, and Semple had been fully identified with it.

But *Geography of the Mediterranean Region* was an impressively comprehensive and erudite work of scholarship, particularly in the eyes of academics in the adjacent fields of classics, sociology, and history, and it was an enjoyable, accessible read.[2] One contemporary reviewer called it "perhaps the greatest study of the Mediterranean region ever written" (Becker 290). The book's review history indicates that even beyond its appreciative specialized audiences, there was a receptive set of nonspecialist readers: soon after its publication, reviews began appearing on both sides of the Atlantic in mainstream publications as diverse as the *Times Literary Supplement* and the *Spectator* in Britain and the *New Republic* in America. Even the American Jesuitical weekly magazine *America* (which has been called the PBS for Catholics) reviewed the book soon after publication.[3] The "attractive fashion" of Semple's writing (*Spectator*) no doubt helped to make it suitable for "the general reader at leisure in his armchair" (R.P.L.).[4]

It is likely that H.D. was among these readers.Her lifelong companion and lover Bryher owned Semple's *Mediterranean Region* (Smyers 24), and its publication date of 1931 suggests an immediate relevance to H.D., since it appeared just before her 1932 Hellenic cruise with her daughter Perdita and Alice Modern, which became the subject of her story "Aegina" (Silverstein "Planting the Seeds"; Guest 205). Because Bryher supplied many of the historical and academic texts in preparation for the journey that she and H.D. took to Egypt with H.D.'s mother in 1922, it seems highly plausible that Bryher would have done so with this Hellenic cruise. Though the book has been catalogued by literary historians as a part of Bryher's library, it would have resided in H.D. and Bryher's joint library at Kenwin, the Swiss Bauhaus–inspired home that Bryher built for them in 1930–31. [5] Kenwin served as a "bolt-hole," a home base away from family surveillance and a "point of departure for travel" for Bryher, H.D., and their immediate circle (Guest 202). Biographer Barbara Guest has taken great pains in praising the house's "splendid library," containing novels as well as "Bryher's necessary history books," which would have included Semple's work. Indeed, Kenwin became the seat of Bryher's "educational program with Perdita as experiment-subject," as Guest has described it (203), which created the occasion for the wider circulation and dissemination within their family unit of the various elements of the Kenwin library. Semple's engagement with ancient classical texts in *Mediterranean Region* would have been particularly appealing to H.D., given the poet's classicist and Hellenistic interests, as would the historical orientation of all of Semple's geographical work. Semple's sensibilities seem uncannily aligned with H.D.'s in this fashion; in desiring to make her work capacious and elastic, Semple conceived of her work as mined from the same Hellenic vein that interested H.D.; in an April 2, 1911, letter to John Keltie, she describes her 1911 *Influences of Geographic Environment* as "Hellenic in form, but Darwinian in method" (qtd. in Keighren 39).[6]

A historicist approach to H.D.'s relationship to broader ideological and cultural patterns of thought and their modes of circulation and dissemination reveals patterns of repeated exposure to environmental determinist thinking for which Semple's work is most visibly known. These patterns can be brought to bear on *Trilogy*, H.D.'s war poem of the early 1940s, written in London as it was under siege from German bombing raids. As many scholars have demonstrated, this impressive visionary war epic operates as a gathering point for many of the historical and mythical materials that interested H.D. over the course of her career, layered together through the montage-like strategy of the palimpsest that H.D. had deployed in earlier work. She defines the palimpsest as a "parchment from which one writing has been erased to make room for another,"[7] and it appears in *Trilogy* as an explicit metaphor for the poet's attempts to strip away sediments of history to locate and recuperate images of female power, to search "the old highways / for the true-rune, the right-spell, / recover old values" (*CP2* 511). Feminist scholars have discussed H.D.'s palimpsest in temporal terms regarding its intervention into patriarchal history. Reading H.D.'s palimpsest spatially, however, allows us to consider elements of *Trilogy* in relation to dominant forms of geographic epistemology that circulated in the early decades of the twentieth century and to understand the contradictory and ultimately irreconcilable impulses toward the transnational other that H.D. engineers through her use of space.

This chapter adds a new dimension to our understanding of *Trilogy's* archive by examining its strategic deployment of environmental determinist principles, both the insistence on the influence of climate on human culture and social behavior and the underlying formal logic of regional comparison. My argument is not that H.D. was an environmental determinist through and through, but that she integrates aspects of her own personal experience of a war-torn environment in London and her love of Egypt with environmental determinist principles, thereby providing her with the authorial grounding to use the palimpsest as an ordering principle that brings together England and colonial Egypt. The transnational politics of this element of *Trilogy's* poetics are neither wholly orientalist nor completely free of this darker side of difference discourse, though taken as a whole they unsettle ideas of ontological fixity on which national hierarchies are founded.

To examine *Trilogy's* transnational connections in this regard complicates recent critical constructions of England as a "shrinking island" in late modernism of the 1930s and 1940s. The single-totality logic of the British Empire brought with it a crisis in part-to-whole relations, of exactly what was and was not British under the sign of the British Empire. Extraterritorial localities distant from England were nominally a part of a unified body politic, but their inclusion also created a sense of disjunction in terms of what an essentially "British" racial identity might be, not to mention a disjunction in how the British system might be a unified system if some of its major transactions were generated from the far colonies. As Jed Esty has discussed, out of this definitional crisis grew a compensatory need to establish a more localized and purified sense of Britishness by rerouting "big" notions attached to cultural identity and imperial territory into self-conscious and self-critical attention to the "small" routines of domestic daily life. For late modernist British literature of the 1930s and 1940s, as Esty argues, this brand of British particularism constituted an "Anglocentric culture paradigm" (2) in which England viewed itself through an anthropological lens. For expatriate American poets such as H.D., the prospect of engaging in the Anglocentric "inward turn" of the 1930s and early 1940s may not have been a straightforward option. Though compelling, Esty's model is incompatible with many expatriate writers living in England whose artistic output participated in its cultural production of the period. The poem *Trilogy,* which navigates H.D.'s Americanness, England, and Egypt, among other materials, creates something of a generative form of cultural, national, and geographic vertigo.

H.D.'s experimental fiction of the 1920s suggests that she had come into contact early in her career with the discourse of environmental determinism if not with Semple's famous *Influences of Geographic Environment.*[8] H.D.'s 1921 roman à clef, *Paint It Today,* for instance, displays starkly obvious investments in environmental determinist principles through its portrayal of its central protagonist, Midget, who is indelibly formed by the distinctly American environment in which she grew up. Though Midget travels to Europe as a young woman, the natural smells and scenes she encounters in Europe link back in her consciousness to the formative impressions that the natural environment in America exerted on her sense of self, a kind of essential Americanness that she carries with her elsewhere, as Annette Debo has argued about the novel.[9] *Paint It Today,* in fact, articulates a formula that is almost an exact replica of the environmental determinist statements we find in Semple's work that stress the conditioning effects of environment—particularly the effects of climate—on the nature of human society. *Paint It Today* declares, "Language and tradition do not make a people, but the heat that presses on them, the cold that baffles them, the alternating lengths of night and day" (20). In keeping with this, the novel repeatedly situates Midget in relation to American geography or features of the natural landscape as the crucible for constructing her essential American identity. This kind of logic resonates with the beginning assertions of Semple's *Influences of Geographic Environment* that "Man is a product of the earth's surface" and even more potently echoes Semple's imagery of the human embodiments of these environmental influences: "[The earth] has entered into his bone and tissue, into his mind and soul" (1).

Bryher's own writing from the 1920s betrays similar familiarity with environmental determinism. In a sense she was an aspiring academic geographer in her own right. Bryher had always been interested in geography as a material practice and as a way to organize her thoughts and feelings. After all, in the opening to her 1937 essay "Paris, 1900," Bryher proclaims, "All my life I have suffered from 'geographical emotions.' Cities are so much easier to understand than people" (33). Her *Picture Geography for Little Children,* published in 1925 by Jonathan Cape, was dedicated to H.D.'s daughter whom Bryher had adopted and was helping to raise. The copy circulated widely within the family, even crossing the Atlantic, judging by the copy in the Beinecke Library with a handwritten dedication to H.D.'s mother, whom they had nicknamed "Horse": "Horse / From / Bryher / (Trusting it will profit!) / May, 1925."[10] Bryher's *Picture Geography* sought to make global geography intelligible and interesting to children by introducing them to the animals indigenous to each country, teaching them along the way each region's climatic features, agricultural or mineral products, and notable aesthetic practices and/or social customs. Afghanistan is among those regions profiled, which in Bryher's hands is

> a very wild country. It is surrounded by mountains. It is very hard to go to Afghanistan. The Afghans do not want strangers to visit their country. And it is difficult to climb across the mountains. The Afghans like to fight. They are very strong, for the winters are cold and the wind that blows across the passes is so fierce that it can almost knock over a man. Very little is known about the country, for very few people have been able to visit it. For many months of the year the snow on the passes is so deep that no horses can travel on the narrow tracks. (*Picture Geography* 47)

Although much of the book avoids outright environmentalist assertions, the entry on Afghanistan is one of several passages that echo Semple's environmental influence thesis and the moral topography of racial difference that often accompanies it.

Texts in the field of classics with which H.D. was familiar reinforced these ideas and provide further evidence to the claims in chapter 1 that the new geography which Semple had ushered in for English-speaking audiences in the first two decades of the twentieth century or so was not exactly "old" geography during the 1930s. H.D. owned a copy of classical historian Eric Warmington's 1934 book, *Greek Geography,*[11] which selected and translated classical Greek literature to reveal the evolution and development of Greek geographical knowledge and to make the case that the "achievements of ancient Greek exploration were greater than ancient literature records or implies" (xlvii). These achievements, Warmington argued, mean that the Greeks merit consideration as "the creators of geography as a science," despite their obvious technical limitations (xlvii). He characterizes the field of historical geography in which his project operates with the language of environmentalist discourse: "On the facts of physical geography follow the distribution of plants, animals, and then man" (x). The structure of the book itself bears out the logic that geographical foundation affects human results, with four parts devoted, in this order, to cosmology; climatology, physical and political geography; exploration and growth of geographical knowledge; and human development of the technologies of mathematical geography and cartography. In Warmington's book conveying classical Greek geographical ideas, H.D. would have seen the familiar environmental determinist narrative structure discussed in chapter 1: landscape first, human activity as a response second. The content of the classical selections Warmington chooses and translates aligns with environmental determinist ideas as well, especially regarding Strabo, the ancient Greek philosopher and historian often considered to be the first western geographer and whose work H.D. knew and read (Gregory 66). At one point Warmington sets out to "defend" Strabo from those who might see in his work the most hardline sort of environmental determinist sentiments, and he argues that his work displays a "softer" environmentalist model of influence of the sort that Semple advocated in her work, rather than outright environmental causation. Relaying Strabo's ideas about the distinctiveness of continents as geographical units, he breaks off with this parenthetical editorial note: "Strabo holds that the character of a country is due to circumstances and chance; civilization is influenced, not caused, by climate; it is institutions and education which have caused the Athenians and others to be what they are; breeding besides locality produces good domestic animals" (Warmington 34). Ironically, what he then goes on to report from Strabo seems of the most causal environmentalist order; in fact, it could almost be a page out of the book of the early twentieth-century "anthropometric" environmental determinism of Ellsworth Huntington or similar formulations that appeared in the *National Geographic Magazine* that took the measure of racial groups through climate: "In approving of such a division of the continents, he [Strabo] gives as an example the fact that the Indians differ from the Ethiopians in Libya, the Indians being better grown men and less parched by the dryness of their climate" (34).

Through Warmington's hefty selections by Hippocrates of Cos, whom twentieth-century geographers have called an "early environmentalist" (Kish 45), H.D. could observe these ideas about climate, race, and culture being reinforced. Warmington includes selections from Hippocrates's *On Airs, Waters, and Places* under the telling heading "Effect of Climate and Soil on Man." Within the selection, we find deterministic statements that proclaim Asia to be "civilized" and its inhabitants "milder and more patient of toil" because of the "temperature of the seasons—Asia lies towards the east midway between the two risings of the sun, and is situated farther from the cold [than Europe is]" (Warmington 54). Other less localized observations equate temperateness of climate and sociability of human society:

> People who inhabit a country which is mountainous, rough, high, and watery, and where the changes which they experience in the seasons vary greatly, naturally possess a physical form which is large and well-fitted by nature of hardiness and manliness. Savagery too and brute ferocity are possessed by such natures most of all. But those who inhabit

> regions which are depressed, covered with meadows, and stiflingly hot, and experience a greater share of warm than of cold winds, and use warm waters, could not well be large or straight as a rod, but are bred with a tendency to be broad and fleshy, black-haired, too, and swarthy rather than pale. They suffer from excess of phlegm rather than of bile. Manliness and hardiness would not alike be inherent in their souls, though the additional effect of institutions might produce this. (58)

A few lines later, Hippocrates summarizes the important role of environment in shaping culture with this aphorism: "All the other things produced in the soil follow on the nature of the soil" (59). Although these ideas are expressed through classical sources rather than contemporary ones, Warmington's selections and the way they are packaged are informed by the intellectual climate of his present.[12]

H.D. draws from this sort of linkage between environmental influence on the one hand and cultural identity and human behavior on the other in her World War II poem *Trilogy*. This tie between the landscape and human society hinges upon several key ideological underpinnings in academic geography which underwrite the poem's ability to create two kinds of communal affiliations. First, *Trilogy*'s appeals to the "man-land" equation help to create a sense of community in war-torn London in ways that acknowledge, but overcome, the potentially divisive questions of national difference that might distance the expatriate American poet-speaker from the British civilian public. Sarah Graham suggests that H.D. addresses these problems of authority and coalition building by appealing to the power of the artist to speak about weighty matters like war and by imagining a hostile public within *Trilogy* that she can thus address and disarm (172–73). I would submit that a turn to the ideology of environmental determinism helps H.D. to create her standing in the community in canny, subtle ways that are key to her ability to generate precisely that defense of poetry. If the poet-speaker in *Trilogy* is of the landscape, then she is of the culture.

This environmental determinist equation is what makes it possible for the poem to make transcultural and transnational connections between London and Karnak that obviate, at least to a certain degree, ethical questions of access that stem from Egypt's status as a British colony. Moreover, the comparativist logic that comes with this mode of thinking is what allows the poem to construct a transitive relationship between London and Karnak that becomes a source of spiritual uplift necessary in a time of war. If H.D. turns to

Egypt as a new monumental past for western culture, to cite Marsha Bryant and Mary Ann Eaverly's recent work on H.D. and archaeologist James Henry Breasted's "Egypto-modernism," then environmental determinism generated by academic geography enables H.D. to make that turn. Her engagement with one academic discipline is aided by the other. H.D. deploys the apparatus of environmental determinism in both instances of transnational crossing that we find in "The Walls Do Not Fall," but also moves past it because of what her use of the disjunctive formal structure of montage does to geographical space in the poem. H.D.'s experimental poetics unsettle geographical ontology in ways that ultimately allow language itself, "the Word" to use the poem's own terms (*CP2* 519), to organize spatial attachments and ideas of belonging. Like language itself, the transnational politics that this activates are transcendent but also slippery and unstable, shuttling between undermining national hierarchies and reinforcing them.

I focus extended attention on the beginning sections of "The Walls Do Not Fall," the first of *Trilogy's* three poems, because the repeated appeals to environmental influence or environmental pressures help to broker H.D.'s merger with a British public and to launch her use of Egyptian historical and mythological materials as an appealing civilizational and spiritual salve. "The Walls Do Not Fall" opens, significantly, not with the individual poet or her "companions of the flame" (*CP2* 521) but with the landscape of bombed out London. Although this captures the sense of shock and disorientation that war produces, it also follows the environmentalist style of placing the physical environment first and people second. This set of images allows the poem to register wartime destruction in ways that treat what happens to the landscape as a primary motor or foreground at the same time that it highlights the question of who can claim a social connection to that landscape: "An incident here and there, / and rails gone (for guns) / from your (and my) old town square" (*CP2* 509). The possessive pronouns instantly foreground issues of propriety, ownership, and connection. Although there is a sense of joint ownership between the poet and her companions, the two sets of possessive pronouns differentiate an assumed "native" sense of ownership felt by the British citizenry from that of the poet, a figure for the expatriate H.D. whose sense of Americanness she retained despite her transnational movements.[13] Indeed, the sequential ordering and the use of the parenthetical grant ownership first and foremost to the *not-poet* "your," which subordinates the poet's own parenthetical expression of ownership and turns it into a tentative and secondary development.[14] What feminist critics for many years have taken to be a spiritual community of women brought together in this passage as "companions of the flame" of World War II's destruction is not necessarily an easy or stable sort of unity between the poet and her companions/readers.[15] We can, in fact, locate tensions between the speaker and her community, which can be linked to broader anxieties in *Trilogy* about the poet's own abilities to be understood and that are held over from H.D.'s crisis of authority as a noncombatant during World War I and fear of being misunderstood by her public (see Graham 171–72).[16] At the same time, however, the parenthetical insert works as an interrupting gesture that interjects the poet's own right to proprietorship, which starts to pave the way for the creation of an organic community that can include both the poet and her "companions of the flame."

"The Walls Do Not Fall" shores up this gap by repeatedly staging dynamics of external, environmental forces impinging upon the poet and her companions. Some of this is couched in explicitly architectural terms and some of it in a generalized idea of climate and landscape. The poem focuses explicit attention on the environmental effects of war: "over us, Apocryphal fire, / under us, the earth sway, dip of a floor, / slope of a pavement" (*CP2* 510). The nature of London's environment at a time of war is surely man-made in ways quite different from the natural effects of topography and climate that interest academic geographers such as Semple. But H.D. repeatedly conveys the destruction of war through geological details that allow the poem's bank of images to tap into broader sets of assumptions about the relationship between the landscape and human society. The "slow flow of terrible lava" and the "crack of volcanic fissure" are things that the poet and her companions "know" in ways that exert "pressure on heart, lungs, the brain / about to burst its brittle case" (*CP2* 510). In this regard, Semple's own invitation to consider her work as suggestive rather than definitive seems instructive for understanding an adaptable and portable environmentalist rhetoric that we find in H.D.'s hands. Semple claims that she has "purposely avoided definitions, formulas, and the enunciation of hard-and-fast rules" and explains that, after all, "[i]t is unwise to put tight clothes on a growing child" (*Influences* vii).

Environmental influence becomes central to building a sense of community for the speaker and her companions. She and they know each other through "secret symbols" that are recognized through "subtle appraisement" that takes place in silence or in the briefest of greetings (*CP2* 519–20). But these conditions of quiet, secret community building depend, significantly, on an earlier assertion that "[p]eril, strangely encountered, strangely endured" which they encounter in the war-torn environment is something that "marks us" (520). From this perspective, the often-cited articulation of a feminist community that the poem offers, "we nameless initiates, / born of one mother, // companions / of the flame," can be read as an expression of contingency. If we read H.D.'s end stop after "mother" as turning what follows it into further elaboration, then they are "born of one mother" precisely because they are "companions / *of* the flame." The familial nature of community is enabled through shared conditionings of the landscape.

There are, of course, many other images of ingestion and expulsion in which the poet searches for the limits of her own boundaries and what she can take in ("infinity" prompts the poet to "sense my own limit, / my shell-jaws snap shut" [*CP2* 513]). But even in these moments, the speaker's ability to patrol her own boundaries is a response to relentless environmental influence: "I know the pull / of the tide, the lull / as well as the moon" and the moon's "cold immortality," which is what propels her belief that the "whale / can not digest me" (513–14). H.D. does not stage this environment as completely deterministic, nor does she rely upon environmental determinism exclusively. In the Jonah-and-the-whale image that follows the poet's consideration of the moon's immortality, the explicit and antagonistic relationship between self and environment is one that the poet has some ability to resist, albeit passively, as material that cannot be consumed by the environment: "the shark-jaws / of outer circumstance // will spit you forth: be indigestible, hard, ungiving" (514). At other moments, such environmental pressures in fact allow the poet to triumph, to "beget, self-out-of-self, // selfless, / that pearl-of-great-price" (514). But even here this process of empowerment and self-begetting is the result of the effects of environmental pressure that the poet cannot escape.

One of the many sources of spiritual uplift and of an affirming power of perseverance in *Trilogy* comes from its turn to Egyptian materials. As several scholars have suggested, H.D.'s view of Egypt generally challenges other modernists' investments in Greece, advanced by poets such as Yeats, Pound, and Eliot, and departs from T. E. Hulme's binaristic distinction between the hard qualities of Egypt and the soft vitality of Greek culture.[17] Marsha Bryant and Mary Ann Eaverly chart H.D.'s turn to Egypt as a new "monumental past" for the West (435) that challenges "traditional narratives of Western culture by making Egypt its source and spiritual center" (450). This turn to Egypt was spearheaded, they note, by developments in academic Egyptology by figures such as Henry James Breasted and Arthur Wiegall. Semple, too, participates in this project of shifting the focus away from assuming Greece to be an isolated, autochthonous culture that alone launched western civilized culture, although not by reversing the binary by privileging Egypt over Greece but by returning early twentieth-century modernity to the ancients' sense of a capacious and "cosmopolitan" Mediterranean basin (*Mediterranean Region* 10) that posited active interplay between Greece and Egypt. Semple approaches the Mediterranean as the ancients did, defining the Mediterranean as composed of three fronts: Asiatic, European, and African (which Semple quickly clarifies means Egypt). Africa, according to Semple, "has made meager contribution to Mediterranean life and civilization. It has been a poor mediator, limited in its power either to give or receive" (*Mediterranean Region* 6), but Egypt "is a striking contrast" to the rest of Africa. At the very minimum, it held central importance in the Mediterranean due to its fertile soil and irrigation technologies that made it "the granary for the populous Mediterranean countries which were rather scantily provided by nature with good wheat land," which in turn made Roman imperial politicians set their sights on exerting control over the region (98). But while Europe eventually evolves into a "dispenser of gifts" of its own, she notes that it first appeared as a "borrower" from Asia and Egypt (5). Both Breasted and Semple thus reject reigning scholarly sensibilities in the field of classics that by the early twentieth century had reduced its spatial focus, treating Athens as though it existed in a spatial vacuum and was the only game in town (Ruprecht 251). Semple, though, posits a more dynamic model of cultural exchange than Breasted, one that is perhaps more consistent with H.D.'s syncretic palimpsest of Egyptian and Greek materials, such as the chain of "Amen-Ra, / Amen, Aries, the Ram" that *Trilogy* forges (*CP2* 527).

Although H.D.'s interest in Egyptian materials is clear, there is more we might do to understand the techniques H.D. uses to facilitate the incorporation of Egyptian materials into *Trilogy* specifically around questions of ethical access. These questions of access punctuate "Secret Name: Excavator's Egypt," one of the novellas in her 1926 trilogy *Palimpsest,* set in British colonial Egypt. The story in various ways interrogates whether and for what purpose Egypt can or should exist as the possession of foreign excavators, whether archaeological or otherwise. The American heroine of the story, Helen Fairwood, who is an Egyptologist of a sort who travels in Egypt while on holiday, visits the tomb of King Amenophis II. Upon encountering a "crouching Arab" at the entrance of the tomb, Fairwood searches "frantically for baksheesh," a combination of tip and alms, which she offers out of this mental calculation: "If she could get him baksheesh all would evolve itself, be right." She wonders, "Had god admitted her? Was it things like this, signs like this, like Charon and his toll, that theoretically bound us?" (*Palimpsest* 182). As Meredith Miller has pointed out, this act of propitiation does not take place without dehumanizing the Arab, who produces a "hypnotic siss-siss of some beatific serpent" (*Palimpsest* 182) and whom Fairwood compares to Charon in order to domesticate him into a more "familiar" Greek mythological narrative. In giving him baksheesh, Fairwood makes the Arab into what she calls her "psychic guarantee" (182), which for Miller means that he is turned into a "symbol of her own right of entry" (90). I would emphasize the fact, however, that Fairwood perceives this as an "unrequired toll" that she opts to pay voluntarily, and given the series of questions she asks about whether God has in fact admitted her, she seems far from presuming her own right of entry.

In this respect, H.D.'s novella participates in a much larger set of debates surrounding the 1922 discovery and excavation of King Tutankhamun's tomb regarding the right of British archaeologists Howard Carter and Lord Carnarvon to control the excavation and the cataloguing of objects. This debate unfolded alongside a parallel set of tensions over the right of reporting over the proceedings; Carter had signed exclusive reporting rights to the London *Times,* which meant that Egyptian news outlets, much to their chagrin, were dependent upon British news organizations to report events of monumental significance on their own soil.[18] On her trip with Bryher to Luxor and Karnak, H.D. saw the new discoveries of King Tut's tomb being excavated, carried out of the tomb, and catalogued by the British-led archaeological team,[19] and it would have been almost impossible for her not to have been aware of the debates regarding propriety over the tomb's excavation and the rights to report on it given that she and Bryher took their tea in a hotel overrun by many of the journalists covering the event.[20]

In order to authorize both material and linguistic connections to Egypt, "The Walls Do Not Fall" appeals to the comparativist techniques of environmental determinism that fuel the poem's constructions of a palimpsestic layering of disparate western and Egyptian materials. In effect, H.D. is able to orchestrate landscape transitivity between London and Karnak by positing similitude between their architectures and, surprisingly, between their climates as well. In the process, the poem authorizes the speaker's access to Egyptian spiritual regeneration and feminist inspiration through making her as much a subject of its landscape as she is of England's own territory. H.D.'s imagery resonates with the logical extension that results from environmental determinism's man-land equation, that if all culture is influenced by geographic conditions, then vastly distant regions that share common cultural conditions produce the same human responses; even seemingly different cultures, if subjected to similar environmental pressures, will share cultural characteristics that those pressures helped to shape. Semple's *Influences of Geographic Environment,* in fact, articulated this logic overtly by stressing the commonality of people if given similar geographic conditions, a commonality that overrides the nature of race and ideas of racial or national difference. In her preface, she embraces all that environmental equivalencies implied about the irrelevance of national identity and ethnicity; she desires

> to compare typical people of all races and all stages of cultural development, living under similar geographic conditions. If these peoples of different ethnic stocks but similar environments manifested similar or related social, economic, or historical development, it was reasonable to infer that such similarities were due to environment and not to race. Thus, by extensive comparison, the race factor in these problems of two unknown quantities was eliminated for certain large classes of social and historical phenomena. (*Influences* vii)

Semple offers this statement partly because she wishes to differentiate her own approach to anthropogeography from the work of her mentor, Friedrich Ratzel, who relied heavily upon Herbert Spencer's Social Darwinist theories of racial superiority and inferiority that were produced by the "survival of the fittest." She notes that "the organic theory of society and state" is prominent in Ratzel because of Spencer, but also that this theory is "now generally abandoned by sociologists" and as a consequence "had to be eliminated from any restatement of Ratzel's system" (*Influences* vii). Against this element of Ratzel's program, Semple articulates her comparativist approach, which distinctly undermines the significance of race and nation and embraces the possibilities that similar geographic conditions can produce similar outcomes, wherever they are found, and that transplanted populations adapt to the will of new environments regardless of their racial or national origins. For academic geographers, the identification of particular environmental zones that could produce particular cultural effects depended upon the very fact of differentiation and comparison. It is this comparative impulse that enables H.D.'s poetic transregional and transnational linkages and makes them intelligible to her audience.

These transitive properties of geography mean that it is precisely the poet's conditioning to the ruination of the London landscape, one sort of man-land relationship, that allows the poet to fully access the landscape of Karnak. It is, after all, "ruin" that opens the London landscape as well as "the tomb, the temple" to the poet (*CP2* 509). This serves a strategic purpose: if being conditioned by a particular environment makes you *of* that environment, then by experiencing the same destruction as Karnak's environment, the poet grants herself access to all which that environment produced or represents. Indeed, the conditions of "mist and mist-grey, no colour" which follow the colon at the end of the line "your (and my) old town square" seem at first to be an elaboration on the eerie similarity between England's typically moist climate and the misty haze produced by bomb dust and smoke from the many fires that such bombings ignited. Its inclusion in the stanza about Karnak, however, suggests at the level of poetic form that it could potentially apply to the dusty haze of that Egyptian space as well. Even the seemingly straightforward exhortation that the poem seems to make in commanding the poet to enter an "open" temple complex at Karnak betrays a more complex navigation of how the poet relates to space: "there, as here, ruin opens / the tomb, the temple; enter, / there, as here, there are no doors" (*CP2* 509). Scholars have taken these lines as intrinsic justification of the poet's right to enter, for they not only render Karnak accessible to the poet and by extension to the West but they nullify any ethical breach by commanding that the poet must enter. The lack of a seal or boundaries to keep out the British tourist, or the Anglo-western reader, makes the presence of citizens of the West less a violation than invited and justified entry. But what is significant is that this command to "enter," syntactically isolated with a semicolon and a comma at the end of a line, does not emanate from "the tomb, the temple" of Karnak directly or alone but instead is produced by the slide between environmental conditions in both locales.

The homology of environment between London and Karnak is accomplished through sight and sound, signaled by the near rhyme of the poem's opening stanzas. The play on the visual similarity between "here" and "there" and their aural nearness accomplishes a direct form of geographic transitivity that, while maintaining the distinctiveness of the two locales, facilitates their connectedness

> An incident *here* and *there*
> and rails gone (for guns)
> from your (and my) old town *square*:

mist and mist grey, no colour,
still the Luxor bee, chick, and *hare*
pursue unalterable purpose

in green, rose-red, lapis;
they continue to prophesy
from the stone papyrus:

there, as *here*, ruin opens
the tomb, the temple; *enter*,
there, as *here*, *there* are no *doors*;

the shrine lies open to the sky,
the rain falls, *here, there*,
sand drifts; eternity *endures*:

ruin *everywhere* . . . (*CP2* 509, emphasis mine)

In her overview of *Trilogy's* poetics, Alicia Ostriker argues that H.D. creates a "unified web" of various linked sounds that envelop the reader's senses. Among these many recurrent sound patterns, the "here-there" off rhyme, she claims, subtly infiltrates the poem and gradually gains in intensity.[21] We hear *here, there, square, hare, enter, doors, endure, everywhere,* which for Ostriker provides an "abstract ground-tone" upon which the other sounds of the poem build or against which the other sounds stand out; the "here/there" sound creates a casual and spatial feel leading to the desolation that only later becomes fully clear in the poem (342). As much as it sets the emotional tone, the controlling *here/there* aural pattern in the opening lyric sequence operates as a poetic correlative to the importance of geographical locations to the poet's psyche in the poem, drawing attention to the significance of *here/there* as markers for London's linkages to Egypt. Though the "dissonance" between the two spaces is distinguishable, the slant rhyme of here/there suggests that London and Karnak can be made to serve as loose correspondents for each other.

Environmental determinism provides the logical structure through which the two locations can be drawn together through environmental similarities, which allows the poet to access all that Egypt promises. And yet H.D.'s poetic strategies take the comparativist logic of environmental determinism to its extreme, by poetically allowing London and Karnak to blend on the space of the page. In this regard, H.D. radically undermines the ontological fixity of geographical space in ways that pose a direct challenge to academic geography's central assumptions, not to mention reigning cultural and political structures of power of the period. H.D.'s adverbial play with "here" and "there" can in this sense work to blur the geographical distinctions between London and Karnak altogether, making it difficult to determine where one location's territory stops and the other's starts. The linkage "there, as here" repeats several times in the poem's opening; however, it is reversed in the final iteration:

the shrine lies open to the sky,
the rain falls, here, there
sand drifts; eternity endures

"Here" and "there," the adverbs that normally function to fix the distance of objects in space, lose their directional power. Their reversal in this stanza suggests not just interchangeability between the two places but that "here" and "there" are not necessarily distinct or distant from each other in space. When we read the line "the rain falls, here, there" together with the next, "sand drifts; eternity endures," rain appears to be falling in London *even as* sand drifts in Karnak. But the unpunctuated line break following "here, there" wreaks semantic havoc with space. The half-meaning that follows from it creates the prospect that rain falls in both spaces at once: the rain falls here and the rain falls there, or that the sand drifts here and the sand drifts there. Through the various syntactical alignments created by H.D.'s parataxis, the poem merges the two landscapes by blurring the quintessential environmental and geological facets normally used to characterize London (rain) and Karnak (sand). Material geography is collapsed, as there is here and here is there, and so too are whatever distinctive geological and climatic conditions by which an academic geographer would normally classify each region. In this sense environmental determinism offers a modernist master narrative that H.D. adopts only partially and strategically in ways that further her poetic project. Environmental determinism depends upon a positivist epistemology—there is a "there" to space that is fixed and knowable, but the question is how to best access it. H.D.'s poetics aim to reject such master narratives and instead question the ontological fixity of space itself.

What impact do these transformations of conventional geographic distance and cultural difference have on the transnational politics of the poem? From one perspective, the landscape transitivity H.D. creates recognizes contemporary Egypt (Karnak is "already" in ruins, in the present era) but evacuates the landscape of contemporary Egyptians, reflecting operations of orientalist discourse at work. In this respect, *Trilogy* perpetuates the vision of Egypt as an empty, timeless space articulated by Romantic poets and eighteenth- and nineteenth-century travel writers. H.D. certainly observed the visible conditions of contemporary Egyptian life during her trip to Karnak and Luxor in 1923, based on the details from the trip recorded in Bryher's diary. One of these sorts of entries reads: "Train left for Cairo about three. Camels. Mud huts. Palms. Natives. Lunch about 4. Arrived Cairo about 6."[22] The contrast between this record of daily Egyptian life and the empty landscape that appears in *Trilogy* illustrates, at least in this instance, Edward Said's claim that to "write about the modern Orient is either to reveal an upsetting demystification of images culled from texts, or to confine oneself to the Orient of which Hugo spoke in his original preface to *Les Orientales*, the Orient as 'image' or 'pensée'" (101). The textual reality of the orientalist literary tradition in *Trilogy*, in other words, dominates over H.D.'s own lived experience. During the war, H.D. reported to her cousin Hattie's husband, Clifford Howard, that she was reading travel literature about Egypt as an escape, and we see that the texts housed in her library portray Egypt as a ruin of the past, stripped of its contemporary realities.[23] H.D. owned copies of both Isabella Romer's two-volume *A Pilgrimage to the Temples and Tombs of Egypt, Nubia, and Palestine in 1845–1846* (1846) and Amelia Edwards's *A Thousand Miles up the Nile* (1891). Despite moments of reticence and self-deprecation, Romer, for example, repeatedly draws attention to the ruinous state of Egyptian monumental architecture that mars the landscape (249, 257), declaring at one point,

> Verily Egypt is the land of tombs! and from the first Pharaoh to the last Memlook, its rulers appear to have thought more of building palaces for the dead, than cities for the living; the proof of which is, that their tombs still endure, while their cities have almost all been swept away—aye, even into oblivion. (92)

In considering the dwellings of everyday Egyptians, Romer's interests might be said to be anti-orientalist, and yet her mixed praise of Egyptian rulers begins and ends with precolonial Egypt, shifting any attention away from the significance of real Egyptian life unfolding in the present-day landscape. There are moments in Romer's travel narrative in which she recognizes the essential role that native Arabic speakers played in her journey (for example, helping her to physically manage walking among fallen pillars and over piles of rubble), but these tend to spark personal gratitude on an individual scale more than recognition of the fuller dimensions of Egyptian culture. H.D. similarly evacuates the landscape completely of the traces of contemporary Egyptians who were at the site all around her during her visit.

In other respects, however, the same sorts of geographically transitive properties I have been tracing thus far also allow H.D. to flip the geopolitical dynamics that normally place England in the position of power over a colonial or semicolonial outpost like Egypt. In a complex rhetorical turn, *Trilogy* uses museum discourse to place London in the position of the objectified cultural artifact that would normally be occupied by Egypt, particularly after the discovery and excavation of King Tut's tomb. While *Trilogy* may not launch an explicit critique of British exploitation of cultural artifacts, it does transpose the imperial center and periphery, registering an awareness of the violent dynamics of cultural appropriation. H.D. utilizes museum language to describe London's bombed out landscape, objectifying home with the same collector's impulses that the excavators of King Tut's tomb had displayed. In the midst of the "Apocryphal fire" that spreads over the speaker and her companions near the beginning of "Walls," they pass from cellar to cellar, looking for protection. As the speaker moves across the landscape, she passes on "to another cellar, to another sliced wall / where poor utensils show / like rare objects in a museum" (*CP2* 510). The conditions of ruin, having reduced the urban landscape to a collection of sliced walls, make the buildings take on a fragmented, alien quality. The image of the "poor utensils" suggests not only that London's most intimate markers of everyday domestic life have been subjected to the collector's objectifying gaze but they are "poor utensils," not the "rare objects" of archaeological significance uncovered at Karnak and Luxor. London has materially taken the place of Egypt in this equation, and it is found wanting.

While these archaeological comparisons partially intervene in the uneven dynamics between England and Egypt of the 1920s that the poem engages, the World War II orientation of "The Walls Do Not Fall" brings it into contact with the contemporary political realities of Egypt that in the 1940s were still entwined with England's priorities, although the hierarchy that structured England and Egypt's relationship during this period haunts the poem in more subtle ways. Although Egypt had achieved increasing measures of independence from British rule in the 1920s, marked in part by England's formal recognition of its independence in 1922 and by becoming a constitutional monarchy in 1923 with the crowning of King Fuad, it was still beholden for many years to the British Empire's financial, political, and military interests in Egypt as well as the Sudan (Botman 29). It was not until Nasser's revolution in 1952 that Egypt achieved complete independence (Albertini and Wirtz 250). In much of the early stages of World War II, England conscripted Egypt into its war effort, using it as a key military outpost where it waged crucial battles against General Rommel's German Afrikacorps. Although the Blitzkrieg bombing raids on England occupy a prominent position in England's own narratives about the war and were also aspects of the war that H.D. experienced while living in a flat in London just blocks away from the battery guns of Hyde Park aimed at incoming German bombers, England's involvement in the war was not limited to its own soil. "The Walls Do Not Fall" registers England and Egypt's co-involvement in the war effort, which becomes a platform from which H.D. recognizes the longer history of European military intervention in Egypt that begins with Napoleon. At the same time, H.D.'s poetics threaten to encircle Egypt in a manner of its own.

On one level, the fulcrum of Germany's aggressions become the basis upon which H.D. constructs affinities between England and Egypt as two countries similarly encircled by European aggression. "The Walls Do Not Fall" creates a sense of connection through the etymological link between "cartouche" and "cartridge" in section 10, launched by her consideration of the threat posed to writing in the face of military aggression. The poet explicitly criticizes the tragic burning of books around her in England, regretting that "our books are a floor / of smouldering ash under our feet" (*CP2* 518). The imagery makes a veiled reference to Nazi book burning of the 1930s as a form of violent censorship and refers to the ongoing destruction of books that the Nazi threat during World War II poses. It is "the most perverse gesture" and stands as evidence of the "meanest of man's mean nature." For the poet, this catastrophe is compounded by the British military's need to commandeer paper for cartridge cases, the outer fittings of individual bullets that contain bullet, primer, and propellant:

> yet give us, they still cry,
> give us books,
>
> folio, manuscript, old parchment
> will do for cartridge cases;
>
> irony is bitter truth
> wrapped up in a little joke,

and Hatshepsut's name is still circled
with what they call the *cartouche*. (*CP2* 518)

Although it is unlikely that bullets in World War II were still made with paper cartridges, H.D. perceived this to be the case, as her June 2, 1943, letter to Viola Baxter Jordan indicates: expressing gratitude for an anthology produced in the United States that Jordan's friend sent her which was "beautifully set-up and printed on such good paper," H.D. contrasts this with the publishing landscape in England, "where so much has been pulped for cartridge cases and all the impediments of munition making."[24] This destruction of the word for the sake of military machinery reflected in part an island nation's acute struggle to produce cheap pulp in a way that could only hope to keep up with the demands of war and yet also sustain writing. This recognition allows the passage to affirm, and yearn for, the long-standing permanence of Egyptian and Greek forms of writing that contrasts with the vulnerable, more ephemeral forms of the West: "Thoth, Hermes, the stylus, / the palette, the pen, the quill endure" (*CP2* 518).

H.D.'s italicization of "cartouche" performs another sort of linkage between aggression and writing: it highlights the long modern history of various forms of Egypt's vulnerability at the hands of European incursions and does this in part at the level of etymology. Like so many of the other moments in which H.D. dismembers words or traces their origins or associations ("Amen" becomes "Amen-Ra," "Osiris" becomes "O-Sirr-Hiss"), the italicization of "cartouche" and its placement in the same section as "cartridge" draws similar kinds of linguistic connections. "Cartridge," the term for a bullet casing, is a corruption of "cartouche," the oblong shape that surrounds Egyptian hieroglyphics, which was a term imposed by the French. It was brought into being by Napoleon's Egyptian expedition of 1798–1801, which literally encircled Egypt as the first form of modern European colonization and subdued the country militarily with one kind of cartouche—the bullet cartridge. The French "encircled" Egyptian culture intellectually by labeling its hieroglyphic expression, like another kind of cartouche, the product of Napoleon's extensive retinue of geographers, botanists, scientists, and naturalists who catalogued Egypt and turned it into an object of French knowledge.[25]

Does H.D. resist these forms of domination or allow them to stand? The transnational politics of this etymologically encrypted formulation of Egypt's experience of European dominance grows more complex because of the montage poetics of the text. The line break and the ambiguous "and" that lead off the lines about Queen Hatshepsut suggest that she, and by extension Egypt, are "still" circled on multiple levels: by the legacies of French intervention which now, in the present moment, bleed into England's encircling of Egypt in order to turn it into the munitions it needs for its World War II efforts.[26] The "and" functions as a simultaneous intervention into these conditions, creating a sense of separation from the World War II violence detailed in the passage such that H.D.'s poem aids in preserving Hatshepsut's name as a form of writing that is encircled, protected. The imagery on this level seems to enact a form of transnational affiliation and sympathy: given the way that England's position in World War II as an island nation under siege made it feel as though it were surrounded by the Germans, the poem creates an analogical relationship between England and Egypt as two cultures, with two forms of writing, both under European threat. The danger is, of course, that Hatshepsut's name and, by extension, forms of Egyptian cultural expression are encircled by the poetics of "The Walls Do Not Fall." H.D. may well recognize these complicated and contradictory conditions in these images of layering and doubleness: "irony is bitter truth / wrapped up in a little joke."

It is significant that H.D. remained steadfastly in London during World War II, despite having had the opportunity to escape to the United States. Nevertheless, she did feel claustrophobic, as many did after the restrictions on food, material goods, and petrol wore away at morale. Writing to Silvia Dobson, H.D. declared yearning for "vast spaces, now that we are so shut-in" (December 1945 letter, qtd. in Dobson 33). Forging extraterritorial connections with Egypt might well have at least begun to address this affective desire for the outside, the elsewhere. For H.D., and for many other expatriate writers like her, the response to Britishness under assault—whether from anxiety about the multivalent ways that capital, culture, and people might flow back and forth between England and its colonies, or from German bombing raids—could not necessarily take the form of a single-national set of affiliations.

The comparative tools of academic geography and its narrative techniques enable this sense of spatial enlargement, although not surprisingly H.D. reserves supreme agency for the genre of poetry and the power of the poet, even over the nature of space itself. She is aware of both time and space as social constructions which she as a poet can form and re-form in her writing. When her literary executor and friend Norman Holmes Pearson asks her to define her poetic practice and chronicle what propels her to write, she deflects his question with one of her own: "Life? Poetry? Times, and places?"

H.D. goes on to reject the familiar elitist critique leveled against poets like her that claim modernist poetry is produced from an insular "Ivory Tower," abstracted from the particulars of life. Instead of directly articulating how her art relates to her social and cultural surroundings, H.D. denaturalizes ideas of time and place, making wordplay of the question itself: "Times, places, dates don't seem so much to matter. Yet there are the times and places of these fragments, as well as I can time and place them" (12 December 1937, qtd. in Hollenberg 8–10). Diana Collecott reads H.D.'s use of aporia and delayed chiasmus here as a strategy to highlight the textuality of language itself (*H.D.* 130). H.D.'s linguistic strategies do more than this, evacuating "time" and "space" as meaningful coordinates in dimensional reality in order to highlight the arbitrary nature of time and space as categories constructed by the human mind. Indeed, H.D. makes the construction of time and space a kind of activity in which the poet, especially, can engage, a process signaled by the change in syntactical function as "time" and "space" transform from nouns to verbs. *Trilogy* seems at moments especially adept at "spacing" in ways that undermine ideas of conventional geographic and cultural fixity. We can view this refusal of firm categorization and compartmentalization as at least some movement toward anticoloniality in the erosion of national and cultural boundaries and the questions of propriety that accompany them.

Conclusion

My study has traced the circulation of ideas about geography in late nineteenth- and early twentieth-century American culture, a span of time marked by a shift from the cosmological geography of European geography well known in nineteenth-century America to the rise of environmental determinist geography in the early twentieth. Although these were two distinct and very different stages, they each adopted in their own way a global orientation and outlook. For Humboldt and Somerville, this meant demonstrating a divinely planned global oneness in which all things are connected and harmoniously ordered. For Semple and Huntington and their middlebrow magazine environmentalist counterparts, the ostensible task was to define the influencing conditions of the geographic environment in particular regions or cultures in the quest for a newly scientific disciplinary model of geography. Implicitly, however, environmental determinist methods depended upon generating global comparisons of local-to-local, national-to-national, or continent-to-continent units, irrespective of geographical propinquity. The global American poetry that I feature here adopts similar methods of comparison as it engages with geographical ideas about the human-landscape equation. Whitman, Hughes, Stein, and H.D., however, take these ideas and methods further. In drawing together disparate global geographies on the space of the page, "presencing" them in new spatial configurations if you will, they each generate a space-collapsing geopoetics that brings into being new transnational and transcultural attachments enabled by these geographical traditions. As the geopoetics of this archive of global American poetry tests the aesthetic and political limits of the day, it also speaks back to geography's fundamental assumptions, undermining the teleological master narratives of environmental determinism as well as the ontological fixity of the institutionalized geographic project. What I have traced, then, is a broad and dynamic conversation about geography and the nature of communal affiliation to which geographers, magazine writers, and modernist poets contributed. In various ways, these three groups weighed in on the nature of the relationship between human populations and the geographic environment, on scientific and aesthetic approaches to space, and on how culture, nation, empire, and race might be conceived or reconceived.

Key to mapping geopoetic formations in these ways is the shared use of formal parataxis to generate the variety of geographical and geopolitical reimaginings I have traced here: for Whitman and Hughes, anaphoric parataxis produces cosmological oneness and a centrality for African diasporic experience, respectively; for Stein, parataxis of syntactic units of language complicates models of nation; for H.D., montage complicates both nation and empire. In reading formal parataxis as performing both geographical and geopolitical work, my project reformulates several familiar ideas about experimental modernism. First, it takes the largely formal notion of "spatial form" generated by Joseph Frank and considers it through a historicized and geopolitical lens. As many modernists may recall, Frank viewed spatial form—that is, an embrace of simultaneity in the act of meaning-making rather than sequence—as the formal disruption of time and narrative. But rather than operate mainly as a formal experiment with epistemology or hermeneutics, we can recognize that spatial form can have direct cultural, national, and/or racial consequences for identity and affiliation, rather than function only to disrupt other formal dimensions of meaning-making.[1] Indeed, the contradictory geopolitics I identify in Stein and H.D. complicate the binary structures of national exceptionalism and imperialism and orientalism traditionally conceived that assume that culturally dominant forces are total and monolithic in their exercise of representational power over the other. The layered and uneven geopolitics that Stein and H.D. ultimately produce also help to dislodge lingering critical commonplaces that still tend to assume avant-garde modernism's aesthetic insularity. The kind of attention in this project to representations of geography in the contexts of dominant geographical epistemologies of the period also deepens our appreciation of the ideological and formal continuities between Harlem Renaissance and Anglo-American modernist texts, cultural and aesthetic formations whose interconnections have been traced with increasing frequency but not, to date, with the robustness they deserve.[2] In the case of Hughes, it creates a more visible place for poetry within narrative-heavy models of the African diaspora and a more expansive internationalist frame for thinking about Hughes's career and affiliations before his so-called radical period of the 1930s.[3]

The intertwined nature of geographical epistemologies and geopoetic practices comes into view when tracing the disciplinary circulations of geography in an American context, a historicizing sort of attention to geography that American Studies scholars have generated about earlier historical periods but that has been late in coming to the modernist period. The national circulation of these ideas, significantly, is not itself determinative of an exclusively American content in the poems under consideration nor the politics that they expressed. That is, the dissemination, uptake, and modification of geographic ideas did not automatically entail the creation of Americanness on the space of the page that just so happens to take place "elsewhere" on the globe. So while I recognize a commonly shared American heritage among the poets I discuss, I have attempted to avoid some of the pitfalls of what Jahan Ramazani calls "culture-of-birth determinism" ("A Transnational Poetics" 343), which dooms particular writers to reproduce their nation of birth in their art. Instead, I follow the global contours of physical migration, reading practices, long-distance and textually mediated affective attachments, and the imagined literary geographies these produced, taking those as the basis for the politics that result.

My project aligns, then, to a certain degree with recent transnational American Studies models that consider America within an expanded spatial arena of interaction, in part out of an effort to avoid reproducing America's exceptionalist dynamics of the past. Border studies models of American studies, for instance, have stressed mutual transformations that emerge when discrete identities come into contact. As John Carlos Rowe characterizes it, this kind of transnational comparativist work in American Studies attends to "the points of historical, geographical, and linguistic contact where two or more communities must negotiate their respective identities" (*New American* 53). Certainly we can see these elements of contact, consistent with a transnational American Studies paradigm, in Stein's representation of the geographical history of America. Here, the intellectual traditions of America in the form of Semple's environmentalist *American History and Its Geographical Conditions* come into contact with Stein's French location and possibly along with it with French geographical models of possibilism developed by Paul Vidal de la Blache and his followers.

This border studies approach to transnational American Studies, however, does not account for the other poetic projects under consideration here. H.D.'s American heritage, for example, comes into contact with England and

Egypt in *Trilogy*, but it is a poem in which H.D. uses American geographic ideas from Semple in order to overcome that Americanness and merge with a British citizenry at a time of war. They also allow her to forge a material connection with Egypt, with the politics of this move informed primarily by British-Egyptian and Franco-Egyptian relations in ways that have little to do with America. Not all of the writers here identify as Americans equally or in the same way, nor do they activate their American-fostered geographic disciplinary heritage in ways that reflect back onto America. But they are united through their reading practices and influences and through the shifting and multiple forms of attachment and belonging that their poetic projects generate. Uniting them all is the shared interest in generating global comparison and exploring new transnational possibilities at the level of content and form.

By positioning poetic modernism, academic geography, and nonacademic geography together as parts of a dynamic terrain of debate, this project also sheds light on some of the blurry spots in our understanding of literary modernism's relationship to science, spots that mirror and are compounded by the ways that geography is positioned today as a discipline. While science studies scholarship in modernism has focused on literary responses to physics, mathematics, and the biological sciences (or their occult alternatives),[4] boundaries drawn around these "hard science" subfields have perhaps unwittingly contributed to the neglect of subfields such as geography that were perceived at the time as "science" but whose methodological and disciplinary makeup included both the natural world and the social. Indeed, in the nineteenth century and for the early twentieth, the label of "science" was often applied both internally and externally to the entire field of geography as a direct result of the comprehensive range of geography's methods and because the category of science itself was more capacious. Today, geography is divided up into the more quantitatively driven subfields of physical geography (geology, hydrology, biology, ecology) and the more qualitative subfields of human geography (cultural geography, historical geography, population geography), and the field as a whole continues to contend with how to make sense of its own internal divisions between "science" and "social science" methods.[5] The field has never really conformed to the "two culture" division between science and the humanities, which C. P. Snow's 1959 influential *The Two Cultures and the Scientific Revolution* has come to represent, with this "two culture" division mapping onto earlier geographic formations in a particularly awkward fashion. Considering the discursive environment in the modernist period in which perceptions of geography's status or profile were shaped frees us of these limitations of the present.[6] It allows us to rediscover in historical academic geography the sorts of relationships between the nonhuman and the human, the scientific and the social, that the "Great Divides" addressed by Bruno Latour and others have artificially suppressed.

Although my discussion here focuses on modernist poetics in relation to cosmological geography and then an environmental determinist paradigm of geography, this method is a generative one capable of opening up fresh possibilities for how we might approach the spatial axis of meaning-making for a wider range of poetic texts and geographic paradigms.

A brief consideration of T. S. Eliot's 1922 *The Waste Land* in relation to environmental determinist ideas, for instance, allows us to notice a more complex relationship between the weather and the natural topography of the poem and its message about civilizational decline.[7] Eliot's poem may not yield as much thick, sustained consideration of landscape-human relations as we find in any of the writers under sustained consideration here due in part to the aural quality of the poem.[8] Whitman, Hughes, Stein, and H.D.'s poems also actively seek to construct spaces of identification and renewal, whereas in Eliot's project this prospect is darker and more tenuous. And yet Eliot's wasted land and its link to civilizational decline[9] take on deeper meaning in relation to environmentalist ideas.

Instead of merely serving as a convenient metaphor for the decay of western civilization (and perhaps the glimmer of hope of renewal), nature goes awry in the poem's world as does the logics of the conventional environmental determinist script. The environmental gifts that western civilization's temperate climate normally provides are out of sync or broken, and whatever the environment does happen to offer, humanity is out of step with it. "April is the cruelest month," because it tries to renew a landscape that is either itself desolate and dysfunctional (its "dull roots") or has been made so by western civilization (Eliot 53).[10] Although this aspect of Europe's climate is performing its typical rainy springtime rituals, its "cruel" impact is the opposite of that of the nurturing Mother Earth celebrated in a text like Semple's *Influences of Geographic Environment*. Winter takes on the wrong seasonal qualities, as it "keeps us warm," the anemic agent "feeding / A little life with dried tubers" (5–6). In part 3, "The Fire Sermon," the "river's tent is broken" (173) and "The wind / Crosses the brown land, unheard" (174–75). The hostile or unavailable natural environment expresses a deeper sense of anguish that whatever traditional environmental gifts the West is supposed to be able to enjoy now fail western culture. The waste in the poem,[11] furthermore, refuses the teleological narrative of progress that environmental determinism generates about the West: if the West is supposed to enjoy the superior gifts that a temperate climate provides, there is also the expectation that these well-equipped cultures are singularly poised to be able to conquer their environments. The overall environment of waste—of ruination or excess—in which the poem's vignettes are situated implies that the West has failed on both of these fronts or, worse, it has been harmed by the excessive environmental gifts of its superior temperate climate or that it no longer deserves to enjoy them.

Historicizing spatial epistemologies and considering their conditions of circulation and production in both geographic and literary texts can also reveal other sorts of convergences between modernist experimentation and geographic models of space and culture. As the purchase of environmental determinism within academic geography grew weaker in the mid-1920s, a more diffuse set of geographic methods and orientations came onto the scene, although two relatively identifiable trajectories can be mapped. A brief consideration of their relationship to modernism indicates a remarkable variety of poetic encounters with major geographic epistemologies of the later modernist moment. With the formation of political geography in the 1920s, which focused on the relationship between formal political structures and space, geopolitics emerged as geographers began to knit their work closely with governmental foreign relations and foreign policy and took an instrumentalist approach to land's political use value. We can see some of its contours animating Ezra Pound's *Pisan Cantos*. The other major geographic development of the mid-1920s, the possibilist ideas about region articulated by Carl Sauer, can be linked quite readily to poems like William Carlos Williams's *Paterson*.

Geopolitics, which eventually became practically synonymous with political geography of the period, stressed the strategic nature of the relationship between geographic space and political power and assumed that spatial enlargement was key to the health and vitality of the nation-state. In a sense, it returned geography as a profession to its nineteenth-century roots, when geographers worked almost as a direct outgrowth of government needs through the production of geological surveys and maps. This paradigm became a significant part of the geographic landscape in America as well as in Europe and Asia. In America, Isaiah Bowman helped to form this direction of the field, returning to the ideas of *lebensraum* (living space) in Friedrich Ratzel's work, which assumed that a political social body was an organic, moving entity that needed room to maintain its vitality.[12] European geographers developed these ideas as well in ways that underpinned nation-state strategies, although in no cases as visibly as Hitler's Nazi regime, which drew heavily upon the work of Karl Haushofer to form its mandate of state enlargement.[13] In principle, Italy's own brand of geopolitics, *spazio vitale,* had a good deal in common with Germany's.[14] *Spazio vitale* referred to the vital space beyond Italy's national borders that was required to ensure the health and vitality of the nation-state; it allowed for some of this space to be occupied by other peoples, but Italy would solely benefit from organizing, administering, and controlling the "vital" territory.

Pound's *Pisan Cantos* of the 1940s formulate a striking literary example of *spazio vitale* that appeared just as Mussolini's actual geopolitical designs had started to crumble and when Pound found himself incarcerated for treason against the American government in the DTC American military camp in Pisa. Throughout his monumental long poem, Pound takes on the voice of an Odysseus poet-wanderer figure, whose vision of the world brings together parts of the historical and literary record as different and as distant as Odysseus's voyage in Homer's epic, the late medieval Italy of Sigismundo Malatesta, various ancient African cities, ancient China, Japan, Jeffersonian America, contemporary Pisa, and more. To a more extreme degree than the other modernist poets I examine here, Pound's method of placing fragments "ply upon ply," as he announces at several turns in the *Cantos,* is a paratactic one, but in this case it is parataxis-as-geopolitics.

The *Pisan Cantos* can certainly be read as a reflection of Pound's sense of crisis given his incarceration, but they also channel his continuing support of Mussolini in geographical and geopolitical terms.[15] A geopoetic approach to the *Pisan Cantos* seems apt, given that this major cluster of the *Cantos* takes the geography of Pisa as its name. The *Cantos* is in many ways not only a "poem including history," the terms Pound laid out for his epic,[16] but also a poem that includes and remodels geography for geopolitical ends, namely, the geography of Italy, Japan, and China.

Like many of the *Pisan Cantos* penned during this period, Canto 74 is a web of snippets of guards' and prisoners' conversations with excerpts from the materials Pound had at hand (a Bible, a *Time* magazine, a volume of Confucius, and a copy of Speare's *Pocket Book of Verse*). It was written at a time when hostilities between Japan and China were again coming to a head. After decades of hostility that were fostered by Japan's incursion into Manchuria in 1894, the second Sino-Japanese war in 1937 had been a significant Asian prelude to World War II. During the war itself, Japan's alliance with Italy under the Tripartite Pact of 1940 bolstered its aggression against China and other Asian countries (Keegan 245).

Certainly, then, the *Pisan Cantos* are situated in a political and military nexus of oppositions: Japan and Italy against China. Yet in Canto 74, distinct markers of the Italian landscape blend with those from Japan and China without allusion to the political fault lines:

> . . . sinceritas
> from the death cells in sight of Mt. Taishan @ Pisa
> as Fujyama at Gardone
> when the cat walked the top bar of the railing
> and the water was still on the West side
> flowing toward the Villa Catullo
> where with sound ever moving
> in diminutive poluphloisboios
> in the stillness outlasting all wars
> "La Donna" said Nicoletti
> "la donna,
> la donna!" (*Cantos* 74/447)

Pound imaginatively transforms a Pisan mountain he sees from his cell into Taishan, the sacred mountain in China upon which many shrines and temples stand (C. Terrell 365). Literally meaning "exalted mountain" in Chinese, Taishan to many Chinese carries with it spiritual symbolism as the "seat of God's authority," for a piece of stone from the mountain is supposed to help ward off evil (Wand 9). Mt. Taishan and the spirituality that the text invests in it become part of a nexus through which several locales and cultures merge. Through the use of the geographically inflected "@" sign, the text maps Mt. Taishan and its spiritual properties directly onto the Italian landscape. The geographical borders separating China and Italy, upon which so much violence and hostility during WWII rests, become textually erased. Similarly, the sacred mountain in Japan's Honshu region, Mt. Fujiyama, becomes mapped onto northern Italy at the Gardone Riviera, a town on Lake Garda where Mussolini established his Salo Republic after being ousted from Rome (C. Terrell 365). Mt. Fuji, whose name is derived from Huchi or Fuchi, the Aino (I-no) Goddess of Fire, and which means "never dying," is still climbed as a part of an old Shinto custom as a means of gaining spiritual purity (F. Davis 79, 131). Its spiritual properties are conferred on the Gardone Riviera by Pound's textual practices.

This elegiac effort to merge the spaces of China and Japan with Italy as a source of spiritual uplift for Italian culture and also for Pound is also a geopolitical project. Merging Fujiyama and Pisa poetically reconstructs Mussolini's Tripartite Alliance between Japan and Italy, in a way that compensates for Mussolini's downfall but that comes at the price of an orientalist lumping of Asian cultures. As Pound breaks down the geographical boundaries that separate the countries for the sake of their symbolic and cultural similarities, he removes the possibility of national difference upon which the political violence of the war rests. Sacrificing the cultural specificity of each location as well as its historical context allows for the fluid transmission of culture that would break down national differences and move beyond the violence of border aggression and the contest for a dominant place in a global hierarchy.

And yet the price of creating this utopic new geography is that the text makes Asia into a cultural monolith, ignoring the contours that differentiate Japanese and Chinese spiritual practices and symbolism. The simile brings Mt. Taishan and Mt. Fuji close to each other, syntactically blurring the geographical and cultural boundaries that separate them *within* the East. Functionally equivalent, Mt. Taishan *becomes* a Pisan mountain just as Mt. Fujiyama relocates to Lake Garda, as both China and Japan, quite interchangeable, offer up the same kind of spiritual salve for Italy. The geopoetics of *The Cantos* thus re-creates on the page the forms of spatial enlargement for Italy that Mussolini envisioned through *spazio vitale*, only here with a compensatory widening of the spatial aperture beyond the Mediterranean and as far as East Asia. This gesture entails subsuming Asian locales into an Italian geography without any ethical consideration of the political and material conditions that would complicate such connections.

The other, and quite different, major paradigm to emerge in the 1920s was the possibilist model of regionalism, which turned toward more purposefully empiricist methods in an attempt to redefine the field as a social science. Carl Sauer's influential analysis of areal differentiation attracted geographers to the study of various forms of urban, regional, and environmental development and management (Alan Baker 74). The possibilist orientation of Sauer's approach was shaped in part by his return to the early anthropological contributions of Franz Boas to formulate a model of human-environment interactions that gave the human side of the equation more agency. Although there were still elements of environmental determinism lurking within this strain of academic production, geographers like Sauer stressed the human ability to modify physical environments (Peet 328). Environment played a role, but human society exerted its own agency as well. This regionalist studies tradition of geography dates back to Strabo,[17] and we can see these traditions surfacing in Sauer's work of the 1920s, whose breakthrough "Morphology of Landscape" argued for the importance of making regional distinctions. Sauer recognized the role of environmental influence articulated by Semple and others but argued, significantly, for human society's ability to adjust in response to both social and environmental factors and that the "interrelation" (299) between human and environment is a context in which we can see "man expressing his place in nature as a distinct agent of modification" (307).

The scope and driving force behind William Carlos Williams's long poem *Paterson* (published in five volumes between 1946 and 1958) can be understood in relation to Sauer's work, and viewing it in this light helps to bring into sharper focus the ways that Williams's project departs so starkly in its geopolitical sensibilities from Pound's. *Paterson* performs a sort of geologic coring of Paterson, New Jersey, that allows it to reflect in a vertical sense on the deep history of the region as well as of the United States, as its geologic layers reveal connections to larger scales and to a wider national space. But by looking inward and downward in a way that embraces particularity ("no ideas but in things"), Williams avoids conscripting space or culture into an instrumentalized superhighway of cultural or national domination.

The poem has traditionally been viewed as exploring the relationship between poetry and prose, the potential for language to represent the real, the inauthenticity of mass culture (Billitteri 59), and the conditions for invoking the cultural authority of the poet (Anthony Flinn), but most scholars on some level recognize, as Christopher MacGowan does, that the poem aims to "bring a wider context to bear upon local history" (49). MacGowan's characterization that in the poem "consequences are determined by prevailing conditions as much as by the pressure of the past, and these conditions are contingent upon factors that can be redirected" (49) seems like it could just as easily be an articulation of Sauer's possibilitist approach to regionalism.

Volume 1 of *Paterson* opens with an image of a man-city of Paterson, New Jersey, a sleeping, larger-than-life being who embodies both the hydrological and topographical features of this region in New Jersey and distinctly human features: "Paterson lies in the valley under the Passaic Falls / its spent waters forming the outline of his back" with the crashing thunder of the falls "filling his dreams!"(6). While initially this imagery seems to subsume the human into the landscape, or indeed stage a scene of geographic invasion into the human, gradual separation emerges across the poem between the man and the city, signaled among other things by the shift from this form of geographical personification to Williams's increasing use of a simile: he becomes a man "like a city" (7).

Rather than the environment shaping the man-city in a deterministic fashion, Williams's increasing reliance upon similes allows Paterson the man not to mirror the environmental features of the city's region but to chart his own course: "Jostled as are the waters approaching / the brink, his thoughts / interlace, repel and cut under, / rise rock-thwarted and turn aside / but forever strain forward" (7–8).[18] Although he's "rock-thwarted" in the most seemingly concrete and immobilizing of ways, he can "strain forward," movement that is emphasized by the lateral and vertical zigzags of his thoughts. This is not only physical and mental movement, but also the signal of the poet's own ability to "strain forward" in piecing together new poetic strains that weave together colonial history, documentary materials such as letters, and mass cultural material (such as the *National Geographic Magazine* and its clichéd titillating function) in ways that allow them to retain, insistently and democratically, their particularity.

Although brief, this consideration of Pound and Williams begins to gesture toward the broader expanse of terrain on which key geographic epistemologies circulated through geographic and experimental poetic texts. For these two poets, as for others, geography can be seen as lending or enabling formal structures, ideas, and forms of cultural authority that are useful to, or provide a call to action in, modernist poetry. Examining modernist experimental poetry and geography together helps to give a new habitation and a name to modernism's spatial, cultural, and geopolitical orientations, its new languages, and its new forms.

Notes

Introduction. Geographical Encounters, Modernist Geopoetics

1. For example, Jon Hegglund's *World Views* (21-23) and Marjorie Howe (in "Joyce, Colonialism, and Nationalism" but particularly in "'Goodbye Ireland'") have focused on geography in *Portrait of the Artist as a Young Man,* although their attention has focused primarily on its relationship to imperialism and colonialism. Approaches to geography and cartography through the lens of Empire are established in Joyce studies, as Hegglund's treatment of *Ulysses* in his "Hard Facts" suggests, although scholars have not examined Joyce's relationship to environmental determinist paradigms within the discipline of geography itself.

2. The publication's original title was *The National Geographic Magazine.* Then, in the 1950s, it became the more familiar *National Geographic.* Throughout my discussion I will be using the former, as it was the title in use during the modernist era.

3. Here I borrow Tim Cresswell's sense of *anachorism,* the term for disordered space spatially equivalent to anachronism's sense of disordered time (103).

4. This title was first codified at the inaugural conference of the Modernist Studies Association, October 7–10, 1999, appropriately called "The New Modernisms: The Inaugural Conference of the Modernist Studies Association."

5. In American Studies, transnational work extends back to the early 1990s. For early formulations, see Wald, "Minefields"; and Porter. For recent transnational, postnational, or planetary approaches to American Studies, see Radway; Rowe, *New American* and *Post-Nationalist;* and Dimock. For an overview of the transnational turn in Modernist Studies, see Walkowitz and Mao's essay "The New Modernist Studies."

6. For work on modernist transnationalism, see work by Timothy Brennan; Melba Cuddy-Keane; Jessica Berman; Jed Esty; Robert Livingston; Michael Moses; Michael Moses and Richard Begam; and Rebecca Walkowitz. Ramazani's "Poetry, Modernity, and Globalization" discusses the narrative-heavy focus of much of the recent scholarship

on globalism. Although these studies variously approach the global through the lens of cosmopolitanism, transnationalism, imperialism, and globalization, I use the language of local, transnational, and global because these terms are more traditional to the academic and popular geographic contexts I consider and because they highlight issues important in this project regarding spatial scale, spatial mapping, and ideas of propinquity more so than discourses of cosmopolitanism.

7. Consider the telling example Ramazani points out in the form of Martha Nussbaum's reference to the "narrative imagination" of cosmopolitanism in her 1997 book *Cultivating Humanity* (290).

8. For discussion of the need for historicist approaches to transnational formations, see Bill Maxwell's response to Ramazani in Wai Chee Dimock's special issue of *American Literary History* devoted to transnational citizenship.

9. Alan Golding has observed that recent Americanist work prefers prose forms, particularly fiction, thanks to the assumption that they are more intimately engaged with the sociopolitical "real world" (xiii). Joseph Harrington traces this through the drama of institutional history, observing that "the assumption that we go to novels to find historical reality because novelists represent historical reality reifies field boundaries by producing a dichotomy between prose narrative as the bearer of historical value, on the one hand, and poetry as the repository of aesthetic value, on the other" (509).

10. Morris provides a longer genealogy of approaches to the cultural engagement of poetic form itself (4–7). For others who formulate such cultural engagements, also see Davidson; Nelson; and Watten.

11. Rachel Blau DuPlessis calls this the "extractive attitude" of much poetry criticism (*Genders, Races* 7). I build upon DuPlessis's concept of "social philology" meant to counter this, a strategy that traces the genealogy of the poetic utterance to broader historical concerns and back again so as to recognize it as an intervention in larger social formations in the world (*Genders, Races* 11). We can find similar formulations in Jerome McGann's idea of "radial reading," strategies that involve consulting the historical and geographical sources to which the details in the text refer that are embodied in "the figure of a person who rises from reading a book in order to look up the meaning of a word in a dictionary or to check some historical or geographical reference" (27). These approaches seem implicitly responsive to Hans Robert Jauss's idea of the "initial horizons of expectations" that the writer's own reading public carries with it (23).

12. Fleming's puckish rewriting of Stephen's list into verses seems inspired by the same kind of boredom or frustration expressed by figures like Germaine Greer, who once said, "The world's a wonderful place; how can geography teachers make it so boring?" (qtd. in Koelsch 261) or by Joseph Conrad's essay "Geography and Some Explorers" in which the academic geography of Conrad's youth and those who taught it are a "a bloodless thing, with a dry skin covering a repulsive armature of uninteresting bones" (254). Even dismissive sentiments about geography as an academic subject and its related pedagogies can, as Conrad's essay attests, fuel creative projects.

13. Here I refer to Raymond Williams's well-known essay "When Was Modernism?" Brooker and Thacker's reformulated version of it privileges geographic location (3).

14. For examples of locational feminist theorizations and approaches, see Frankenberg and Mani; Friedman, *Mappings*; and Walsh.

15. Both Hegglund (*World Views*) and GoGwilt provide valuable analysis of geography's contribution to geopolitics as a strategy of territorial consolidation or enlargement, although they address Anglo-European modernist fiction, memoir, and film, and do not discuss the role of environmental determinism in any significant degree. Jessica Berman's chapter on American geography and Stein in *Modernist Fiction, Cosmopolitanism, and the Politics of Community* is a partial exception, although its examination of Stein's *The Geographical History of America,* which I discuss in chapter 4, serves a larger analysis of Stein's novel *The Making of Americans.*

16. Narrative-focused scholarship on geography in earlier historical periods of American literature and culture can be found in the work of Anne Baker; Bauer; Brückner; and Brückner and Hsu. One of Hsu's chapters in *Geography and the Production of Space* notably devotes sustained attention to Whitman, although most of the book focuses on narrative.

17. See David Harvey, "Cosmopolitanism" 532–35 for discussion of the environmental determinist elements of Kant's *Geography,* and Edward Casey's response, which stresses Kant's avoidance of a priori historical conditions and notes his shift away from such geographical equations in general in later work.

18. A fuller discussion of the ways that environmental determinism drew upon social Darwinism can be found in Peet and also in Willinsky's chapter, "Geographies of Difference," in *Learning to Divide the World: Education at Empire's End.*

19. For a discussion of the poem's gender and sexual dynamics, layered upon ideas of racial otherness, see Edelman; and Bowen. See Hawkins for a discussion of the poem's attempts to construct cross-ethnic and cross-national oneness.

20. For more information on the ways that environmental determinism was used to justify colonialism, see Peet. Also see Natter and Jones 196.

21. Historical geographers and historians have gestured toward the ways that environmental determinism lingered beyond the 1920s, though they have not explored this afterlife in much detail. See G. Martin *All Possible*; Livingstone; and Schulten.

22. See, e.g., the new feature "Historical Geography at Large" of the *Journal of Historical Geography,* which journal editor Felix Driver considers a platform for examining the public forms of geographical knowledge. Driver and Schwartz have encouraged similar forms of scholarship in *History Workshop Journal,* which moves across disciplinary boundaries as well as those that might separate academic from public forms of cultural practice. Cavell's recent *Tracing the Connected Narrative: Arctic Exploration in British Print Culture, 1818–1860* exemplifies this kind of attention to the crossover between academic and popular sites of geographical knowledge production. Also see Driver, "Geography's Empire" 35. Historians like Schulten and Rothenberg have recently be-

gun to contribute to understanding the popular cultural terrain on which geographic knowledge has taken shape.

Chapter 1. Academic and Popular Geography: Global Connections, Environmentalist Style

1. See Kadlec for a brief discussion of the Lamarckian elements of Oppenheim's essay, which he situates in relation to melting pot theories of race (288n18).

2. For discussions of early colonial nationalism in relationship to geographic texts, see Brückner; and Brückner and Hsu. For discussions of anxieties about expanding borders and racial inclusionism in the mid- to late nineteenth century, see Anne Baker.

3. Schulten, e.g., describes a "pre-professional stage" of the nineteenth century in which geography was dominated by scientists and engineers working for the federal government (46). One possible reason for this is that the preoccupation in early American geography with land use and with natural topographical features reflected a nation that operated through rural units, a nation that produced geographers who were born and raised in small towns and whose orbit of knowledge revolved around the agrarian landscape (Mikesell 11).

4. Under Benton's guidance, John C. Frémont led the U.S. Army Corps of Topographical Engineers in their comprehensive mapping of western territories (Allen 63–64). Also see Koelsch 248.

5. I differentiate between "scientist" and "geographer," even though these were not meaningful distinctions in the early to mid-nineteenth century, when geography was itself a "science" that contained other scientific fields such as astronomy, physics, botany, and the like.

6. American cartography throughout most of the nineteenth century produced domestic national maps and atlases; it was only beginning in the 1870s, the golden age of American cartography, in which global mapmaking became a mass industry that made maps of the world affordable and more commonplace (Schulten 17). Recognizing the circulation in America of cosmological academic geography originating in Europe allows us to notice earlier forms through which the globe came to American notice.

7. Emerson's claim is quoted in Laura Dassow Walls, *The Passage to Cosmos* (252). Aaron Sachs describes Humboldt's stature relative to Einstein's; also see Pandora 352. For Humboldt's influence on other American Renaissance writers such as Thoreau, see Walls's *Seeing New Worlds* 76–166.

8. This is because of Humboldt's reputation and also because geology was thought to be the proper repository of physical geography. But some geographical historians have come to question this disparity in reputation; Somerville's *Physical Geography* was actually begun in 1841 and completed in April 1842, before Humboldt's work appeared (J. Baker 207), and other historiographers have suggested that Humboldt's *Cosmos* was influenced directly by Somerville's own earlier works (Neeley).

9. Blanchard and Lea, e.g., published the 1850 edition of *Physical Geography*. For

discussions of American reprinting of "foreign" texts, see Loughran; McGill; and Wadsworth. As this body of scholarship suggests, the term "reprinting" may well be more appropriate than "piracy" in an age before more rigid copyright law. I limit my use of the term "piracy" to discussions of Somerville's own objections to seeing her work published without her consent and without monetary reward.

10. Somerville writes in her memoir that *Physical Geography* went through nine editions and was translated into German and Italian and that the book "went through various editions in the United States, to the honour, but not to the profit, of the author. However, the publisher obligingly sent me a copy" (*Queen of Science* 162).

11. For just some of the reviews of Somerville's *Physical Geography*, see Brewster ; "Editor's Book Table" 1848; "Editor's Book Table" 1850; "Literary Notices"; "Review of *Physical Geography*, revised London edition."

12. Both Schulten and Brückner discuss the relationship between geography readers and national identity.

13. According to Peet, determinism "attempted to explain the imperial events of late nineteenth and early twentieth century capitalism in a scientific way," and in the process, "geography had also to legitimate intersocietal competition and the conquest of some societies by others" (310).

14. For a discussion of these comments made by Mackinder to the Royal Geographical Society, see Schulten 72.

15. As most accounts of environmental determinism would have it, American geographic discourse maintained an "umbilical" connection to the earth's physical features (Smith 183).

16. Davis conceived of ontography as a geographic springboard for human development, postulating the "inorganic element of the earth . . . acting as a control . . . and some element of the existence or growth or behavior or distribution of the earth's organic inhabitants, serving as a response" (G. Martin "Ontography" 279).

17. David Livingstone is not unique in proclaiming American geography in the early decades of the twentieth century, at rock bottom, a "Lamarckian-derived attempt to couch geography's traditional concerns in the language of evolution" during a period of increasing professionalism of the field and growth of specializations within it (212). Also see Schulten 78.

18. See Keighren for a discussion of Brigham's work (33) and an overview of Semple's relationship to Ratzel (40). Isaiah Bowman's early career also advanced environmentalist approaches to region, thanks to the influence of William Morris Davis, with whom he studied, although he shifted noticeably from environmentalist work to geopolitics in the 1920s. For a discussion of Bowman's career and trajectory, see Smith 47.

19. See, e.g., Everett Edwards's obituary for the journal *Agricultural History,* which in 1933 was claiming the continued centrality of Semple's *American History and Its Geographic Conditions,* "still a standard reference for students of American history and geography" (150).

20. Schulten argues that part of environmental determinism's appeal might lie precisely in the ways that it did not emerge in a systematized, unified way (indeed, the unifying label "environmental determinism" is one that geographers applied retroactively and pejoratively). It was articulated by multiple academic geographers and in slightly different form in Frederick Jackson Turner's frontier thesis. But the appeal of Semple across generic, stylistic, and disciplinary boundaries might be an equally compelling reason.

21. For discussions of imperialism and environmental determinism, see Peet.

22. See Lutz and Collins's groundbreaking study of the *National Geographic Magazine* for a discussion of the nature of the "middlebrow" periodical and a critique of the ideological messages the magazine's articles and photo layouts conveyed.

23. For a discussion of the *National Geographic* as a "cultural standard-bearer" and its relationship to taste and class, see Rothenberg 21–22. See also Lutz and Collins; Hawkins.

24. Hawkins is an exception—see, e.g., her consideration of Hemingway. But it is worth noting that neither Hawkins's book on the *National Geographic* nor any of the other recent monographs on the periodical have been reviewed by academic journals with a modernist focus.

25. Flanner, e.g., viewed the assignments offered by the *National Geographic* as a much-needed form of escape from New York conservatism. See W. Murray 38, 41.

26. Willis's survey of Moore's wider reading habits provides some sense of the company the *National Geographic* kept in American readers' personal libraries; it catalogs the *Literary Digest* (which excerpted articles from literary, political, scientific, and religious magazines), the Christian journal *Outlook*, the *Illustrated London News*, the *English Review* (ed. Ford Maddox Ford), *Scribner's Magazine*, and the *Boston Evening Transcript* (Willis 7).

27. Mabel Dodge Luhan Papers, Beinecke Rare Book and Manuscript Library, Yale University.

28. Hodgson Papers, Beinecke Rare Book and Manuscript Library, Yale University.

29. June 19, 1964, James Weldon Johnson Collection, Beinecke Rare Book and Manuscript Library, Yale University. Permission to reprint granted by the Langston Hughes Estate, Harold Ober Associates, agents.

30. Affordable volumes were "Tauchnitz and the Conrad, and they didn't go much further than Kipling and Hardy in those days. Our moderns, particularly when pounds and dollars were translated into francs, were luxuries the French and my Left Bankers were not able to afford. That is why I was interested in my lending library" (Beach 21).

31. Although there is not a complete inventory for all of the years of the lending library's operations, a 1944 inventory of the bookstore lists several issues from the 1930s as in circulation (Oct 1939, Oct 1938, Jan 1932, Dec 1936) and several listed as not having arrived yet (March 1940, Sept 1938, Sept 1939, June 1936). Sylvia Beach Papers, Harvey S. Firestone Library, Princeton University.

32. Letter from H.D. to Bryher, September 3, 1935, Beinecke Rare Book and Manuscript Library, Yale University (© 2014 by The Schaffner Family Foundation, used by permission of New Directions Publishing agents).

33. Although Schulten does not discuss Semple's involvement in the National Geographic Society, she notes the continued involvement of William Morris Davis, even after he himself formed the American Association of Geographers, and complicates the idea of a schism advanced by Pauly, Koelsch, and others. By addressing this interplay between academic and popular forms of geographic knowledge, Schulten's work valuably expands the scope of historical geography beyond a narrow, "internalist" approach that considers institutional contexts alone. This expanded scope of her historical research allows her to trace the presence of environmental determinist discourse in the *National Geographic* in light of the national demands presented by the Spanish-American War of 1898.

34. See Poole for a brief mention of the environmental determinist qualities of the magazine, and Schulten for more extended discussion focusing on the years surrounding the Spanish-American War. This historical moment "encouraged those at the *National Geographic* to think about historical development in light of this relationship between natural environment and human behavior. This formulation enabled the *Geographic* to valorize the nation's work abroad, which would aid those less able to master their own environment" (Schulten 52). See Tuason for a similar discussion. The fact that environmental determinist discourse appears in the periodical before the Spanish-American War, however, can be attributed to the fluid interplay between academic and popular forms of geography. For a discussion of the ways that the *National Geographic* operated as a handmaiden to American nationalism and American exceptionalism, see Rothenberg; Lutz and Collins; Poole. Hawkins examines reader resistance to the sometimes unevenly expressed nationalist concerns within the magazine.

35. Most scholars have focused on the cultural work of the photographs in the magazine and the implied power of the photograph to objectify and capture the cultural other in various states of "primitive" display (Lutz and Collins; Willinsky) or the tensions between stylized representation and scientific objectivity (Hawkins).

36. Discussions of the environmentalist formal sequence often found in geography textbooks appear in Peet (328) and Wrigley (7).

37. The Royal Geographic Society initially launched the short-lived *Geographical Magazine* in 1874, soon to be replaced by the *Proceedings of the Royal Geographical Society*. This academically minded periodical folded in 1892, but its academic orientation informed *Geographical Magazine,* a new magazine launched in 1935 that is now called *Geographical* (Johnston 93).

38. Hammerton's *Countries of the World* series was published by Harmsworth's Amalgamated Press, for which Hammerton worked for a number of years and which at the time was the largest publishing empire in the world (Hadaway).

39. Hammerton subsequently republished *Countries of the World* as bound illustrated books, a move that is strikingly similar to the *National Geographic*'s generation of coffee table books that Stephanie Hawkins describes, the first one of which, *Scenes from Every Land,* was published in 1913 (69).

Chapter 2. The "Terraqueous" Globe: Walt Whitman and the Cosmological Geography of Humboldt and Somerville

1. For an overview, see Brückner and Hsu 12.

2. For a discussion of the "international" quality that non-American readers have long seen in Whitman's work, see Grünzweig, "'Collaborators'"; and Thomas.

3. Arthur Golden exemplifies this tendency to dismiss Whitman's later work, especially "Passage to India." For reevaluations of this tendency, see in particular Warren.

4. A range of scholars have discussed the poem's relationship to these exclusionary practices, in particular Grünzweig, "Imperialism"; Hsu; Paryz; and Wertheimer.

5. Dressman discusses Whitman's use of Samuel Goodrich's 1855 *The World as It Is, and as It Has Been; or, A Comprehensive Geography and History Ancient and Modern* when working on the 1860 *Leaves of Grass.*

6. For scholars who have briefly noted Humboldt's influence on Whitman in relation to his earlier work, most often his poem "Kosmos," see Thomas 97–98; Reynolds 245; Scheick; and Matteson. Walls is an exception of sorts, devoting a section of her *Passage to Cosmos* to Whitman, although it is fairly brief and focuses on the environmental ethics important to Humboldt that she claims "Passage to India" misses in its support of manifest destiny. Her work offers valuable recognition of the extensive Humboldtian influence on multiple figures of the American Literary Renaissance whom Whitman knew, including Emerson, Thoreau, and Poe.

7. Whitman also invokes the "kosmos" in the preface to the 1855 *Leaves of Grass,* where its capaciousness allows him to portray the poet as a democratic equalizer and unifier during the buildup of tensions in the 1850s regarding slavery.

8. See "Passage to India" in Michael Moon's Norton Critical Edition of *Walt Whitman: Leaves of Grass and Other Writings* (345–53). Subsequent parenthetical references to this poem and Whitman's "Kosmos" will refer to line numbers within the Moon edition. Whitman completed "Passage to India" in April 1870. It was later featured as the first poem in an annex of the 1871–72 *Leaves of Grass* (Reynolds 499).

9. Trent Collection of Whitmaniana, David M. Rubenstein Rare Book and Manuscript Library, Duke University. Although unlabeled, the pagination and appearance of Whitman's clipped review match the original entry in the *North British Review.* The pagination differs in subsequent bundled volumes of multiple *North British Review* issues, as well as in further reprintings of the review in the American *Eclectic Magazine of Foreign Literature,* and perhaps others.

10. See Houghton for identification of Brewster's authorship (1: 671). The *North British Review* was a relatively new magazine at the time, created to provide a more politically progressive alternative to the well-established *Edinburgh Review.* The *Edinburgh Review* contained fewer reviews in order to generate longer and more in-depth coverage of only the most interesting or groundbreaking new publications; the *North British Review* took this as its formula. Its early circulation was as high as four thousand copies sold, exceed-

ing the *Edinburgh Review*, and then settled into a pattern of selling between two and three thousand copies a year (Shattock and Wolff 148; Houghton 1: 663).

11. For example, Whitman handwrote a transcription of an article from an 1844 issue of the *North British Review* on Michaud's *History of the Crusades* ("North British Review, May 1844"), which the Walt Whitman Archive estimates to have been composed in 1869.

12. Hsu in particular reads the poem this way, an approach that is similar to Wertheimer's, which excavates its historical imperialist dimensions in relation to South American contexts. My argument is more in line with Paryz's reading of the poem, which sees its orientalist representations of the non-West ultimately questioned by Whitman.

13. Reynolds discusses Whitman's marginal notations of an 1849 article from the *Edinburgh Review*, "Taylor's Eve of Conquest," which quotes from Humboldt's *Kosmos*: "Humboldt, in his *Kosmos*, citing Schiller, has observed of the Greeks: 'With them the landscape is always the mere background, or a picture, in the foreground of which human figures are moving'" (Reynolds 245, 614).

14. Laura Dassow Walls states that Whitman was "said to have written *Leaves* with a copy of *Cosmos* on his desk, and he took notes on Humboldt and preserved newspaper clippings about him" (*Passage to Cosmos* 280), although Walls does not provide citations for these particular claims.

15. Both Erkkila (*Whitman* 156) and Folsom ("'A spirit'" 598–99) link these images to the political turmoil of the civil war but stop short of considering Whitman's inspiration for selecting them.

16. For discussions of Whitman's identification with Lincoln, see Cavitch; Blasing; Price and Folsom; and Ramazani, *Poetry of Mourning*.

17. The homosocial nature of Whitman's fascination with Humboldt further informs this kind of reading given the connections in Whitman's work between male erotics and brotherhood. For example, see Fone's discussion of the *Calamus Poems* and the brotherhood of lovers (213, 250) and Erkkila's analysis of brotherhood ("Whitman and Political Theory" 137). Other connections to Humboldt might similarly follow fraternal lines. Thomas briefly links Whitman to Humboldt through Whitman's younger brother Jeff, who was a topographical engineer (98).

18. For discussions of Whitman's deep involvement in the printing process, see Folsom, "'A spirit of my own seminal wet'"; and Henry, who details the frequent appearance of printing terms like "justifying" in other work by Whitman.

19. The York Cathedral Pulpit formally censured her for her religious views (Sanderson 418).

20. I refer to Whitman's small notebook containing assorted jottings about "Passage to India," which exists in the Lion collection at the New York Public Library. See Bowers for a transcription and a discussion of the drafts and typescripts of "Passage to India."

21. See, e.g., Gregory Woods's discussion of British modernists who work in a Whitmanian vein.

22. H.D. published this review in *Life and Letters Today* under the pen name–like pen name of "Sylvia Penn."

Chapter 3. African Diasporic Re-Placing: Race and Environment in the Poetry of Helene Johnson and Langston Hughes

1. Hughes wrote the poem "Old Walt" to celebrate the impending one hundredth anniversary of *Leaves of Grass* (Hutchinson, "Langston" 17), which was published in the *Beloit Poetry Journal Chapbook* in 1954, an entire volume dedicated to the poet (Hughes, *Collected Poems* 678n446). He and Arna Bontemps also included several Whitman poems in their 1949 *Poetry of the Negro*, under the header "Tributary Poems by Non-Negroes."

2. For discussion of Hughes's Whitman columns, see Hutchinson, "Langston"; also see Klammer 1–2.

3. For discussion of Whitman's influence on Hughes, see Hoffman; Hutchinson, "Langston"; and Folsom, "So Long, So Long!"

4. I quote Hughes's poetry from *The Collected Poems of Langston Hughes*, indicated in the text as *CP1*.

5. For discussion of these aspects of the poem, see Lynes, "'A Real.'"

6. The representational power of the Anglo-European explorer, naturalist-scientist, geographer, and travel writer has long punctuated geography's history, as Mary Louise Pratt has observed in *Imperial Eyes*.

7. For another reading that opens up a sense of the poem's mediated qualities and its broader geographical orientations, see W. Jason Miller's consideration of Hughes' "river-places" (33) and American lynching culture.

8. Hurston was one among a number of prominent African American intellectuals, writers, and editors who trained as anthropologists, ranging from Arthur Huff Fauset, half-brother of Jessie Redmon Fauset, to Eslanda Robeson.

9. See Livingstone for a discussion of Boas's critique (290–91). For an overview of Boas's relationship to African American racial politics, see L. Baker, *Anthropology*; Lamothe.

10. See the review, "Semple's *Influences of Geographic Environment*," in the *Nation*. Dworkin (636) mentions that Hughes subscribed to this magazine, as did George Schuyler, who was also a contributor (Hutchinson, *Harlem Renaissance* 128).

11. Semple's book appears in the "What to Read" section of the November 1911 *Crisis*.

12. See Livingstone for a discussion of Semple's participation in anthropometric ideas.

13. For discussion of the magazine's signification of cosmopolitan taste and erudite global knowledge throughout the twentieth century as a "high middlebrow" magazine, see Lutz and Collins 7; see also Schulten 153–54.

14. Verner Mitchell notes that this poem was originally published by Johnson with this spelling in the October 1926 issue of *Palms*, but upon subsequent republication (including its appearance in both Hughes and Bontemps's 1949 anthology *The Poetry of the Negro* and Maureen Honey's 1989 *Shadowed Dreams*), the spelling somehow became "Magalu" (12, 13).

15. For discussion of the poem's critique of Euro-American colonialism and Christianity, see DuPlessis, *Genders, Races* 128.

16. Katherine Lynes's essay "'Sprung from American Soil'" considers the poem's engagements with genre as it explores ecocritical terrain, primitivism, and the tensions between nature and poetry.

17. Scholars such as Westover have noted the diasporic qualities of the poem, although its spatial and epistemological dimensions have not been considered.

Chapter 4. (Trans) Nation, Geography, and Genius: Gertrude Stein's *Geographical History of America*

1. For these linguistics terms, see Roman Jakobson. For an overview of Stein's alternating experiments with the axis of selection and the axis of combination, see Cyrena Pondrom's introduction to *Geography and Plays*.

2. Excerpts are from *The Geographical History of America; or, The Relation of Human Nature to the Human Mind* by Gertrude Stein, copyright © 1936 and renewed 1964 by Alice B. Toklas. Used by permission of Random House, an imprint and division of Random House LLC. All rights reserved.

3. Also see Stein's September 20, 1935, letter to Thornton Wilder (Burns and Dydo 55) in which Stein describes not so much having finished the *Geographical History* but having "stopped."

4. Leo Stein, Gertrude's brother, was intensely interested in theories of history (Wagner-Martin 127), which may have brought Stein into contact with Semple's works of historical geography.

5. See, e.g., Ellen Berry's *Curved Thought and Textual Wandering*; Harriet Scott Chessman's *The Public Is Invited to Dance*; Marianne DeKoven's *A Different Language;* Elisabeth Frost's *The Feminist Avant-Garde in American Poetry*; Franziska Gygax's *Gender and Genre in Gertrude Stein*; Bob Perelman's *The Trouble with Genius*; Marjorie Perloff's *The Poetics of Indeterminacy;* and Catharine Stimpson's "Gertrude Stein and the Transposition of Gender."

6. See, e.g., P. Davis. For discussions of Stein's contrast between American wholeness and European imperialism's expanding vastness in *Lucy Church Amiably*, see Ashton.

7. For discussions of Stein's anxieties about her newfound publicity and about her relationship to America as an expatriate, see Goebel; and Vanskike. Steven Meyer is an exception to this tendency, although his approach to *Everybody's Autobiography* focuses on Stein's appeal to scientific theories of memory and to an Emersonian tradition of radical empiricism in order to tap into a "transcendent" notion of American identity that one might conceivably access from anywhere.

8. Vanskike maps this interpretative territory of *Useful Knowledge* with an eye toward the problems of identity that Stein negotiates in *The Geographical History of America*.

9. Goebel (243) and Vanskike (160) discuss the distinctions between human mind and human nature in *Geographical History* with particular care. I am indebted to Vanskike's wonderful reading of *Geographical History,* which attributes Stein's distinctions between the human mind and human nature and her disruption of temporality and chronology to the interrogation of the nature of identity in the text.

10. William Gass declares it unlikely that Stein was familiar with Turner, although he nevertheless sees parallels between his frontier thesis and her focus on the spaciousness of America (11). He and the other scholars who have discussed the *Geographical History*'s engagement with geography have not addressed the ways that Stein might have come into contact with these ideas through this kind of text.

11. Semple's reference to the "central belt" of the North American continent reflects the significant portions of American history that focus on the felicitous nature of America's climate. There are echoes, too, between that set of claims and Stein's attention to weather, as in "There is flat land and weather and money for the human mind" (*GHA* 101).

12. In his letters to Stein, Wilder discussed their friend Aleck Woollcott's radio address for CBS, "Why I Shall Vote for Roosevelt" (September 27, 1936, letter from Wilder to Stein, Burns and Dydo 116), and referred to Roosevelt as a "very great man" (January 1937 letter from Wilder to Stein, Burns and Dydo 127).

13. Burns and Dydo note this sentiment about fathering in their *The Letters of Gertrude Stein and Thornton Wilder* (117).

14. See Jonas for a discussion of Vidal's rediscovery by contemporary geographers working out of a "new regionalism" vein.

Chapter 5. H.D.'s *Trilogy* as Transnational Palimpsest

1. I cite *Trilogy* from H.D.'s *Collected Poems,* ed. Martz. Page numbers of subsequent quotations will be accompanied by *CP2*.

2. Bushong has noted that classical scholars in particular took notice of the work (91).

3. Butler (23) offers this characterization of *America*.

4. Although Keighren's monograph on Semple reassesses the impact of her earlier books, a census of the book review history of *Geography of the Mediterranean Region* reveals a surprisingly wide public readership that would counter any assumption that Semple's career experienced a precipitous decline in the 1920s with the shift within academic geography away from environmental determinism. For some of the reviews of *Geography of the Mediterranean Region,* see "The Geographical Factor"; H.J.F.; Moore; Mortimer; Rev. of *The Geography*; and R.P.L.

5. Zilboorg (xii) points out that most of the books that are housed in H.D.'s library at the Beinecke Library at Yale and Bryher's library in East Hampton, New York, largely constitute these writers' later libraries, leaving a gap in the record about their books from

before the early 1920s. It is tempting to speculate that earlier texts by Semple were among the earlier lost books of H.D. and Bryher.

6. For a discussion of H.D.'s Hellenism, see Gregory.

7. I quote H.D.'s epigraph to her trilogy of novellas, *Palimpsest*. For a discussion of the image of the palimpsest in H.D.'s work generally, its Greek roots, and its relationship to Freudian ideas of projection, repetition compulsion, repression, and trauma, see Friedman, *Psyche Reborn* 28–30. See Riddel for discussion of the ur-patterns of the palimpsest. See Kloepfer for discussion of the palimpsest as a strategy for coming to terms with the figure of the mother; and Collecott, "Images," for discussion of the palimpsest in relation to H.D.'s photomontages.

8. The titles belonging to H.D. and Bryher that we know of today seem to point to many more like them that we do not. H.D. often lent her books to friends and acquaintances, and it is unclear how many of them returned to her possession. Her letter to Ezra Pound indicates lending tendencies that had by the 1930s evolved into something quite systematic and deliberate: "I have been posting books to Bob McA [Robert McAlmon] in Spain on that lending library principle. Let me know if I am likely to have books you might like and I will post to you . . . if you post back . . . not a bad idea" (June 15, 1933, qtd. in Friedman, *Analyzing Freud* 351).

9. Although Debo calls H.D. an environmental determinist in advancing this set of claims about *Paint It Today* (150), her analysis examines the meaning of this term in relation to broader Lamarckian scientific ideas of organic memory theory rather than the disciplinary discourse of academic geography.

10. This is inscribed on the copy of the book located in the Bryher Papers, Beinecke Rare Book and Manuscript Library, Yale University (© 2014 by The Schaffner Family Foundation, used by permission of New Directions Publishing, agents).

11. Smyers lists this among H.D.'s books in the Bryher Library and notes that it bears H.D.'s personal owl bookplate (24).

12. Note that Warmington includes an editorial aside comparing Hippocrates in this vein with Aristotle, who maintains similar theories about climate as determining the nature of human society that dwells within it (59). This suggests that Hippocrates and these elements of Strabo were not all that unusual among the classical authors with whom H.D. was familiar.

13. In "Exile in the American Grain," Friedman discusses H.D.'s Americanness as an enduring part of her particular form of expatriation. Also see Annette Debo's recent and more comprehensive study of the importance of Americanness in H.D.'s literary imagination.

14. My reading of the opening contrasts with Graham's claim that by the end of the first stanza the poem creates an "unusual intimacy" (164).

15. The poem's community of "companions of the flame" has been a central feature in many feminist readings, as we see in Gubar; and Friedman, *Psyche Reborn*. The constructed nature of this formation is not a part of their discussions.

16. For other discussions of the poem's response to the war and to other wartime liter-

ary texts, see Schweik; Edmunds; Friedman, *Psyche Reborn*; and DuPlessis, *H.D.: The Career*.

17. For discussions of the relationship between Greek and Egyptian aesthetics and cultural materials, see M. Miller 96; Bryant and Eaverly 439–40; and Kusch.

18. For information on the controversy about archaeological control, see Hoving 105–6; Reeves 64; and North 19. For discussion of the tension regarding the right to report news of the events reflected in H.D.'s story, see Meredith Miller's essay "Enslaved to Both These Others." For consideration of the story's attitude toward anthropology, see Witte.

19. H.D.'s firsthand witnessing of the spectacles orchestrated by Carter and Canarvon may have informed her awareness of the power dynamics inherent in these practices. The two purposely brought objects out of the tomb, whenever possible, without covers for the special titillation of onlookers (Carter and Mace 144).

20. H.D.'s own visit would have brought her into the closest proximity to these dynamics surrounding the excavation. She repeatedly took tea at Luxor's Winter Palace Hotel (Silverstein, "Planting the Seeds" 10), whose walls were frequently adorned with various postings about Tut's tomb (Reeves 64) and where Lord Carnarvon himself and his family were staying during H.D.'s visit. Howard Carter's January 29, 1923, diary entry reads "Arrived Luxor, Ld. C. put up at Winter Palace."

21. It is useful to note that Ostriker's careful parsing of H.D.'s poetics stands as one example that challenges the recent claims made by Lawrence Rainey in *Institutions of Modernism* that feminist critics, focused only on the content of H.D.'s work, have failed to appraise the skill and public impact of H.D.'s work adequately.

22. Bryher's diary with miscellaneous notes dated 1923–1970, Beinecke Rare Book and Manuscript Library, Yale University (© 2014 by The Schaffner Family Foundation, used by permission of New Directions Publishing, agents).

23. She writes, "I have been keeping warm, reading old Egyptian travel-books, and am so interested in the whole subject of the (roughly) 5000 years BC and all its religious content." Clifford Howard was a fellow writer and married to H.D.'s cousin, Hattie. Letter dated January 1942, Beinecke Rare Book and Manuscript Library, Yale University (© 2014 by The Schaffner Family Foundation, used by permission of New Directions Publishing, agents).

24. Beinecke Rare Book and Manuscript Library, Yale University (© 2014 by The Schaffner Family Foundation, used by permission of New Directions Publishing, agents).

25. See Mary Louise Pratt's discussion of the intellectual forms of domination of such naming and cataloguing projects.

26. Taking a psychoanalytic approach to this passage, Susan Edmunds has argued that H.D.'s imagery replaces the military aggression of war machinery with the spiritual redemption embodied in Hatshepsut's endurance in language after her death. She reads the cartridge/cartouche doubleness as interrogating the nature of divinity and of sexual and gender division, citing H.D.'s awareness of archeological evidence that Hatshepsut pretended to be a man (42).

Conclusion

1. See Frank's *The Idea of Spatial Form,* responses to which have largely focused on the relationship between simultaneity and temporality (see, e.g., Smitten and Daghistany's volume on Frank's theory; and Quinones, *Mapping Literary Modernism*).

2. Houston Baker's *Modernism and the Harlem Renaissance* and Paul Gilroy's *Small Acts,* for example, have pointed out the ways that African American modernism uniquely contended with the history of slavery and institutionalized racism. These specific histories are important to recognize, although they can also inform the ways that we link African American and Anglo-American contributions to modernism in ways that scholars have begun to do.

3. Typically Hughes's internationalist period is identified with the 1930s and early 1940s. See, e.g., Smethurst; and Dawahare.

4. For discussions of modernism and science, see Clarke; Holmes; Steinman; Tichi; and Whitworth, *Einstein's Wake*. For an example of scholarship that addresses the intersections between science and religion in literary modernism, see Vetter.

5. For discussions of the long history of geography's disciplinary identity crisis, see G. Martin *All Possible* 524–27; and Natter and Jones, 196, 202n77.

6. Michael Whitworth has made similar observations about the relationship between physics and modernist poetry ("'Strange Synthetic Perfumes'" 88).

7. It is possible that Eliot was familiar with environmental determinism, given that his job at Highgate Junior School shortly after his marriage to Vivienne included teaching geography (Sanna 19).

8. The poem signals this through its performative and jazz-oriented original title, *He Do the Police in Many Voices*. I find compelling Alan Marshall's view that *The Waste Land*'s city of London is built less by the eye than by the ear, a "place that is made up of sound" (100).

9. Vincent Sherry reads the poem as documenting the British Empire in decline (131), which he roots in part in Eliot's experience working in the Colonial and Foreign Department of Lloyd's bank (115–17). Eliot's sense of decay can be linked to Otto Spengler's *Decline of the West,* as Moretti and others have discussed (223). For Moretti, the poem's use of materials from mass culture to "shore up" the ruins of western civilization limits Eliot to producing something that can be "provisional, unstable, and approximate" at best (210).

10. All references to *The Waste Land* are from the 1963 Harcourt Brace centenary edition; subsequent quotations will be accompanied by line numbers in the text.

11. For a recent discussion of the waste in the poem, see Carver.

12. Bowman published a "minor best-seller," *The New World: Problems in Political Geography,* in 1921 (Smith 182). Bowman began his career as an environmental determinist (he was educated under William Morris Davis just like Ellsworth Huntington), although in the wake of World War I and the work of the Paris Peace Conference in which he participated, he quickly turned to political geography. See also Peet 317–18.

13. Halford Mackinder in England, Karl Haushofer in Germany, and many other European geographers pursued these ideas of territorial expansion as a national imperative.

14. In 1939, for example, Italy formed a new journal, *Geopolitica,* devoted entirely to the topic of Italian geopolitics with Haushofer serving on its editorial board (Klinghoffer 93). For some of the differences between Italy and Germany's versions of geopolitics, see Kallis, *Fascist Ideology* 50–51; and Rodogno, *Fascism's European Empire* 48.

15. See Bush, who has discussed Pound's efforts to preserve Italian cultural identity as an expression of loyalty to Mussolini; also see Nicholls. For discussions of Pound's fascism, which is not the same as geopolitics, see Redman.

16. In his *ABC of Reading,* Pound makes the often-quoted declaration that "An epic is a poem including history" (46).

17. In Renaissance England it developed under the guise of the science of chorography ("place-writing"), a micro-scale approach to geography whose goal was to generate detailed characterizations of small sections of the earth's surface, without necessarily considering the relationship of a particular region to the earth as a whole (Livingstone 76). This intense focus on circumscribed, small-scale regions later underpins historical antiquarians in nineteenth-century New England who sought to study their region's history. For them, history was a vertical series of layers that could be "readable in the strata of both the human-built environment and the workings of earth history" (Halttunen 11).

18. For a discussion of the increasing frequency with which Williams used similes in book 1 of *Paterson,* see Conarroe.

Works Cited

Adams, Harriet Chalmers. "Volcano-Girded Salvador: A Prosperous Central American State with the Densest Rural Population in the Western World." *National Geographic Magazine* 41.2 (Feb. 1922): 189–200. Print.

Adorno, Theodor. "Parataxis: On Hölderlin's Late Poetry." *Notes to Literature*. Vol. 2. Ed. Rolf Tiedemann. Trans. Shierry Weber Nicholsen. New York: Columbia University Press, 1992. 109–49. Print.

Agnew, John, David N. Livingstone, and Alisdair Rogers, eds. *Human Geography: An Essential Anthology*. Malden: Blackwell, 1996. Print.

Ahearn, Barry, ed. *The Correspondence of William Carlos Williams and Louis Zukofsky*. Middletown, CT: Wesleyan University Press, 2003. Print.

Albertini, Rudolf von, and Albert Wirtz. *European Colonial Rule, 1880–1940: The Impact of the West on India, Southeast Asia, and Africa*. Trans. John G. Williamson. Westport: Greenwood, 1982. Print.

Allen, John L. "Working the West: Public Land Policy, Exploration, and the Preacademic Evolution of American Geography." Blouet 57–70. Print.

Anderson, Benedict. *Imagined Communities: Reflections on the Origins and Spread of Nationalism*. London: Verso, 1991. Print.

Ashton, Jennifer. "Gertrude Stein for Anyone." *ELH* 64 (1997): 289–331. Print.

Baigent, Elizabeth. "Holdich, Sir Thomas Hungerford (1843–1929)." *Oxford Dictionary of National Biography*. Oxford: Oxford University Press, 2004. Web. 1 May 2013.

Baikie, James. "The Cradle of Civilization: The Historic Lands along the Euphrates and Tigris Rivers Where Briton is Fighting Turk." *National Geographic Magazine* 29.2 (Feb. 1916): 127–62. Print.

Baker, Alan. *Geography and History: Bridging the Divide*. Cambridge: Cambridge University Press, 2003. Print.

Baker, Anne. *Heartless Immensity: Literature, Culture, and Geography in Antebellum America*. Ann Arbor: University of Michigan Press, 2006. Print.

Baker, Houston A., Jr. *Modernism and the Harlem Renaissance.* Chicago: University of Chicago Press, 1987. Print.

Baker, James. "Mary Somerville and Geography in England." *Geographical Journal* 111.4 (1948): 207–22. *JSTOR.* Web. 7 May 2011.

Baker, Lee D. *Anthropology and the Racial Politics of Culture.* Durham: Duke University Press, 2010. Print.

———. *From Savage to Negro: Anthropology and the Construction of Race, 1896–1954.* Berkeley: University of California Press, 1998. Print.

Bauer, Ralph. *The Cultural Geography of Colonial American Literatures: Empire, Travel, Modernity.* Cambridge: Cambridge University Press, 2003. Print.

Beach, Sylvia. *Shakespeare and Company.* New York: Harcourt Brace, 1959. Print.

Becker, Howard. Rev. *The Geography of the Mediterranean Region: Its Relation to Ancient History,* by Ellen Churchill Semple. *Annals of the American Academy of Political and Social Science* 162 (1932): 290–91. Print.

Begam, Richard, and Michael Valdez Moses, eds. *Modernism and Colonialism: British and Irish Literature, 1899–1939.* Durham: Duke University Press, 2007. Print.

Berdoulay, Vincent R. H. "Geography in France: Context, Practice, and Text." Dunbar 45–78. Print.

Berman, Jessica Schiff. *Modernist Commitments: Ethics, Politics, and Transnational Modernism.* New York: Columbia University Press, 2012. Print.

———. *Modernist Fiction, Cosmopolitanism, and the Politics of Community.* Cambridge: Cambridge University Press, 2001. Print.

Berry, Ellen. *Curved Thought and Textual Wandering: Gertrude Stein's Postmodernism.* Ann Arbor: University of Michigan Press, 1992. Print.

Billitteri, Carla. "William Carlos Williams and the Politics of Form." *Journal of Modern Literature* 30.2 (2007): 42-63. *Project Muse.* Web. 4 Nov. 2013.

Bingham, Millicent Todd, and Raul Blanchard. *Geography of France.* Chicago: Rand McNally, 1919. Print.

Bishop, Elizabeth. *Geography III.* New York: Farrar, Straus and Giroux. 1976. Print.

Blasing, Mutlu Konuk. "Whitman's 'Lilacs' and the Grammars of Time." *PMLA* 97.1 (Jan. 1982): 31–39. Print.

Blouet, Brian W., ed. *The Origins of Academic Geography in the United States.* Hamden: Archon, 1981. Print.

Boas, Franz. "The Study of Geography." *Science* 9 (1887): 137–41. Rpt. in Agnew, Livingstone, and Rogers 173–80.

Botman, Selma. *Egypt from Independence to Revolution, 1919–1952.* Syracuse: Syracuse University Press, 1991. Print.

Bowen, Clare. "Frames of Reference: *Paterson* in "In the Waiting Room." *Twentieth Century Literature* 54.4 (2008): 472–92. Print.

Brennan, Timothy. "Postcolonial Studies and Globalization Theory." *The Postcolonial and the Global.* Ed. Revathi Krishnaswamy and John C. Hawley. Minneapolis: University of Minnesota Press, 2008. 37–53. Print.

[Brewster, Sir David]. Rev. of *Physical Geography,* by Mary Somerville. *North British Review* 11.17 (May 1848): 76–100. Print.

Britzolakis, Christina. "Making Modernism Safe for Democracy: *The Dial* (1920–9)." *The Oxford Critical and Cultural History of Modernist Magazines, Vol. 2 (North America 1894–1960).* Ed. Peter Brooker and Andrew Thacker. Oxford: Oxford University Press, 2012. Print.

Brooker, Peter, and Andrew Thacker, eds. "Introduction." *Geographies of Modernism: Literatures, Cultures, Spaces.* New York: Routledge, 2005. 1–5. Print.

Brückner, Martin. *The Geographic Revolution in Early America.* Chapel Hill: University of North Carolina Press, 2006. Print.

———. "Lessons in Geography: Maps, Spellers, and Other Grammars of Nationalism in the Early Republic." *American Quarterly* 51.2 (1999): 311–43. Print.

Brückner, Martin, and Hsuan Hsu, eds. *American Literary Geographies: Spatial Practice and Cultural Production, 1500–1900.* Newark: University of Delaware Press, 2007. Print.

Bryant, Marsha, and Mary Ann Eaverly. "Egypto-Modernism: James Henry Breasted, H.D., and the New Past." *Modernism/Modernity* 14.3 (2007): 435–54. Print.

Bryher. "Paris, 1900." *Life and Letters Today* 16.8 (1937): 33–42. Print.

———. *A Picture Geography for Little Children. Part One: Asia.* London: Jonathan Cape, 1925. Print.

Burns, Edward M., and Ulla Dydo, eds. *The Letters of Gertrude Stein and Thornton Wilder.* New Haven: Yale University Press, 1996. Print.

Bush, Ronald. *The Genesis of Ezra Pound's "Cantos."* Princeton: Princeton University Press, 1976. Print.

Bushong, Allen D. "Ellen Churchill Semple." *Geographers Biobibliographies Studies* 8 (1984): 87–94. Print.

Butler, Kiera. "Articles of Faith." *Columbia Journalism Review* 44.4 (2005): 22–26. *Communications and Mass Media Complete.* Web. 14 Apr. 2003.

Buttimer, Anne. "Bridging the Americas: Humboldtian Legacies." *Geographical Review* 96.3 (2006): vi–ix. *JSTOR.* Web. 14 Jun 2013.

Carter, Howard. "Diary." *Tutankhamun: Anatomy of an Excavation. Griffith Institute.* Web. 29 Mar. 2013.

Carter, Howard, and A. C. Mace. *The Discovery of the Tomb of Tutankhamun.* Mineola, NY: Dover, 1977. Print.

Carver, Beci. "London as a Waste of Space in Eliot's *The Waste Land.*" *Critical Quarterly* 29.4 (2007): 56–70. *Wiley Online Library.* Web. 7 Mar. 2014.

Casey, Edward S. "Is There Still Room for Freedom? A Commentary on David Harvey's 'Kant's Anthropology and Geography.'" *Reading Kant's Geography.* Ed. Stuart Elden and Eduardo Mendieta. Albany: State University of New York Press, 2011. 285–90. Print.

Cavell, Janice. *Tracing the Connected Narrative: Arctic Exploration in British Print Culture, 1818–1860*. Toronto: University of Toronto Press, 2008. Print.

Cavitch, Max. *American Elegy: The Poetry of Mourning from the Puritans to Whitman*. Minneapolis: University of Minnesota Press, 2007. Print.

Chessman, Harriet Scott. *The Public Is Invited to Dance: Representation, the Body, and Dialogue in Gertrude Stein*. Stanford: Stanford University Press, 1989. Print.

Chinitz, David. "Rejuvenation through Joy: Langston Hughes, Primitivism, and Jazz." *American Literary History* 9.1 (1997): 60–78. Print.

Clarke, Bruce. *Energy Forms: Allegory and Science in the Era of Classical Thermodynamics*. Ann Arbor: University of Michigan Press, 2001. Print.

Collecott, Diana. *H.D. and Sapphic Modernism: 1910–1950*. Cambridge: Cambridge University Press, 1999. Print.

———. "Images at the Crossroads: H.D.'s 'Scrapbook.'" Friedman and DuPlessis 155–84.

Conarroe, Joel. *William Carlos Williams' "Paterson": Language and Landscape*. Philadelphia: University of Pennsylvania Press, 1970. Print.

Conrad, Joseph. "Geography and Some Explorers." *National Geographic Magazine* 45.3 (1924): 239–74. Print.

———. "An Outpost of Progress." *Heart of Darkness and Other Tales*. Ed. Cedric Watts. Oxford: Oxford University Press, 2002. 3-25. Print.

Constable & Co. Advertisement for *The Geography of the Mediterranean Region: Its Relation to Ancient History*, by Ellen Churchill Semple. *Times Literary Supplement* (London), 12 May 1932: 341. *TLS Historical Archive, 1902–2008*. Web. 12 Mar. 2012.

Creese, Mary. *Ladies in the Laboratory? American and British Women in Science, 1800–1900*. Lanham: Scarecrow Press, 1998. Print.

Cresswell, Tim. *Place: A Short Introduction*. New York: Blackwell, 2004. Print.

Cuddy-Keane, Melba. "Modernism, Geopolitics, Globalization." *Modernism/modernity* 10.3 (2003): 539–58. Print.

Darius, Sara. *The Senses of Modernism: Technology, Perception, and Aesthetics*. Ithaca: Cornell University Press, 2002. Print.

Davidson, Michael. *Ghostlier Demarcations: Modern Poetry and the Material World*. Berkeley: University of California Press, 1997. Print.

Davis, F. Hadland. *Myths and Legends of Japan*. London: Harrap, 1912. Print.

Davis, Phoebe Stein. "Subjectivity and the Aesthetics of National Identity in Gertrude Stein's *The Autobiography of Alice B. Toklas*." *Twentieth Century Literature* 45.1 (1999): 18–45. *JSTOR*. Web. 29 Feb. 2008.

Davis, William Morris. "The Improvement of Geographical Teaching." *National Geographic Magazine* 5 (July 1893): 68–75. Print.

Dawahare, Anthony. "Langston Hughes's Racial Poetry and the End of Race." *MELUS* 23.3 (1998): 21–41. Print.

Debo, Annette. *The American H.D.* Iowa City: University of Iowa Press, 2012. Print.

DeKoven, Marianne. *A Different Language: Gertrude Stein's Experimental Writing*. Madison: University of Wisconsin Press, 1989. Print.

Diamond, Jared M. *Guns, Germs, and Steel: The Fates of Human Societies*. New York: Norton, 1999. Print.

Dikshit, R. D. *Geographical Thought: A Contextual History of Ideas*. New Delhi: Prentice-Hall, 1997. Print.

Dimock, Wai Chee. *Through Other Continents: American Literature across Deep Time*. Princeton: Princeton University Press, 2006. Print.

Dobson, Sylvia. "Cloud of Memories: Books H.D. Shared with Me, 1934–1960." *H.D. Newsletter* 1.2 (1987): 26–41. Print.

Doudna, Martin K. "'The Essential Ultimate Me': Whitman's Achievement in 'Passage to India.'" *Walt Whitman Quarterly Review* 2.3 (1984): 1–9. Print.

Doyle, Laura, and Laura Winkiel, eds. *Geomodernisms: Race, Modernism, and Modernity*. Bloomington: Indiana University Press, 2005. Print.

Dressman, Michael. "Goodrich's Geography and Whitman's Place Names." *Walt Whitman Review* 26 (1980): 64–67. Print.

Driver, Felix. "Geography's Empire: Histories of Geographic Knowledge." *Environment and Planning D: Society and Space* 10 (1992): 23–40. Print.

———. "Historical Geography at Large." *Journal of Historical Geography* 36 (2010): 243–44. Print.

Driver, Felix, and Bill Schwartz. "History Workshop Journal: Widening the Circle." *History Workshop Journal* 70 (2010): 1–3. Print.

Dunbar, Gary S., ed. *Geography: Discipline, Profession, and Subject since 1870: An International Survey*. Norwell: Kluwer Academic, 2001. Print.

Dungy, Camille T., ed. *Black Nature: Four Centuries of African American Nature Poetry*. Athens: University of Georgia Press, 2009. Print.

DuPlessis, Rachel Blau. *Genders, Races, and Religious Cultures in Modern American Poetry, 1908–1934*. Cambridge: Cambridge University Press, 2001. Print.

———. *H.D.: The Career of That Struggle*. Bloomington: Indiana University Press, 1986. Print.

Dworkin, Ira. "'Near the Congo': Langston Hughes and the Geopolitics of Internationalist Poetry." *American Literary History* 24.4 (2012): 631–57. *JSTOR*. Web. 13 Jun 2013.

Dydo, Ulla, ed. *Stein Reader: Gertrude Stein*. Evanston: Northwestern University Press, 1993. Print.

Edelman, Lee. "The Geography of Gender: Elizabeth Bishop's "In the Waiting Room." *Contemporary Literature* 26.2 (1985): 179–96. Print.

"Editor's Book Table." Rev. of *Physical Geography*, by Mary Somerville. *Godey's Lady's Book* Nov. 1848: n. pag. *Accessible Archives*. Web. 5 May 2013.

"Editor's Book Table." Rev. of *Physical Geography*, by Mary Somerville. *Godey's Lady's Book* Mar. 1850: n. pag. *Accessible Archives*. Web. 5 May 2013.

Edmunds, Susan. *Out of Line: History, Psychoanalysis, and Montage in H.D.'s Long Poems*. Stanford: Stanford University Press, 1994. Print.

Edwards, Everett E. "Ellen Churchill Semple." *Agricultural History* 7.3 (July 1933): 150–52. Print.

Eliot, T. S. *Collected Poems, 1909–1962*. Centenary Edition. New York: Harcourt Brace, 1963. Print.

English, Hugh. "'By Being Outside of America': Gertrude Stein's 'Geographical History' of Gender, Self, and Writing." *Women, America, and Movement: Narratives of Relocation*. Ed. Susan L. Roberson. Columbia: University of Missouri Press, 1998: 258–80. Print.

"Environment as a Force in Human Action." Rev. of *Influences of the Geography of the Mediterranean Region*, by Ellen Churchill Semple. *Dial* 51 (16 Nov. 1911): 398–99. Print.

Erkkila, Betsy. "Whitman and Political Theory." *Whitman East and West: New Contexts for Reading Walt Whitman*. Ed. Ed Folsom. Iowa City: University of Iowa Press, 2002. 115-44. Print.

———. *Whitman, the Political Poet*. Oxford: Oxford University Press, 1997. Print.

Esty, Jed. *A Shrinking Island: Modernism and National Culture in England*. Princeton: Princeton University Press, 2004. Print.

F., H. J. Rev. of *The Geography of the Mediterranean Region: Its Relation to Ancient History*, by Ellen Churchill Semple. *Man* 33 (1933): 138–39. *JSTOR*. Web. 19 Mar. 2012.

Fitch, Noel Riley. *Sylvia Beach and the Lost Generation*. New York: W. W. Norton, 1983. Print.

Flinn, Anthony. *Approaching Authority: Transpersonal Gestures in the Poetry of Yeats, Eliot, and Williams*. Lewisberg: Bucknell University Press, 1997. Print.

Folsom, Ed. "So Long, So Long! Walt Whitman, Langston Hughes, and the Art of Longing." *Walt Whitman, Where the Future Becomes Present*. Ed. David Haven Blake and Michael Robertson. Iowa City: University of Iowa Press, 2008. 127–43.

———. "'A spirit of my own seminal wet': Spermatoid Design in Walt Whitman's 1860 *Leaves of Grass*." *Huntington Library Quarterly* 73.4 (2010): 585–600. *JSTOR*. Web. 5 Mar. 2014.

Fone, Byrne R. S. *Masculine Landscapes: Walt Whitman and the Homoerotic Text*. Carbondale: Southern Illinois University Press, 1992. Print.

Foucault, Michel. "Of Other Spaces." *Diacritics* 16 (1980): 22–27. Print.

Frank, Joseph. *The Idea of Spatial Form*. New Brunswick: Rutgers University Press, 1991. Print.

Frankenberg, Ruth, and Lata Mani. "Crosscurrents, Crosstalk: Race, 'Postcoloniality' and the Politics of Location." *Feminism and Race*. Ed. Kum-Kum Bhavnani. Oxford: Oxford University Press, 2001. 479–91. Print.

Friedman, Allen. "Stephen Dedalus's *non serviam*: Patriarchal and Performative Failure in *A Portrait of the Artist as a Young Man*." *Joyce Studies Annual* 13 (2002): 64–85. Print.

Friedman, Susan Stanford, ed. *Analyzing Freud: Letters of H.D., Bryher, and Their Circle*. New York: New Directions, 2002. Print.

———. "Exile in the American Grain: H.D.'s Diaspora." *Agenda* 25 (1988): 27–50. Print.

———. *Mappings: Feminism and the Cultural Geographies of Encounter*. Princeton: Princeton University Press, 1998. Print.

———. *Psyche Reborn: The Emergence of H.D.* Bloomington: Indiana University Press, 1981. Print.

Friedman, Susan Stanford, and Rachel Blau DuPlessis, eds. *Signets: Reading H.D.* Madison: University of Wisconsin Press, 1990. Print.

Frost, Elisabeth A. *The Feminist Avant-Garde in American Poetry*. Iowa City: University of Iowa Press, 2003. Print.

Gass, William. "Introduction." *The Geographical History of America; or, The Relation of Human Nature to the Human Mind.* Baltimore: Johns Hopkins University Press, 1995. 3–42. Print.

"The Geographical Factor in American History." Rev. of *The Geography of the Mediterranean Region: Its Relation to Ancient History*, by Ellen Churchill Semple, and *Manual to Accompany American History and Its Geographic Conditions*, by Clarence Fielden Jones. *Geographical Review* 24.1 (1934). *JSTOR*. Web. 19 Mar. 2012.

Rev. of *The Geography of the Mediterranean Region: Its Relation to Ancient History*, by Ellen Churchill Semple. *The Spectator* 149 (20 Aug. 1932). *Proquest*. Web. 27 March 2013.

Gilroy, Paul. *Small Acts: Thoughts on the Politics of Black Cultures*. London: Serpent's Tail, 1994. Print.

Goble, Mark. *Beautiful Circuits: Modernism and the Mediated Life.* New York: Columbia University Press, 2010. Print.

Goebel, Bruce A. "'If Nobody Had to Die': The Problem of Mortality in Gertrude Stein's *The Geographical History of America*." *Philological Quarterly* 70.2 (Spring 1991): 237–52. Print.

GoGwilt, Christopher. *The Fiction of Geopolitics: Afterimages of Culture from Wilkie Collins to Alfred Hitchcock.* Stanford: Stanford University Press, 2000. Print.

Golden, Arthur. "Passage to Less than India: Structure and Meaning in Whitman's 'Passage to India.'" *PMLA* 88.5 (1973): 1095–103. *JSTOR*. Web. 31 Mar. 2006.

Golding, Alan. *From Outlaw to Classic: Canons in American Poetry*. Madison: University of Wisconsin Press, 1995. Print.

Graham, Sarah. "'We have a secret. We are alive': H.D.'s *Trilogy* as a Response to War." *Texas Studies in Literature and Language* 44.2 (2002): 161–210. Print.

Greely, A. W. "Recent Geographic Advances, Especially in Africa." *National Geographic Magazine* 22.4 (April 1911): 383–98. Print.

Gregory, Eileen. *H.D. and Hellenism: Classic Lines*. New York: Cambridge University Press, 1997. Print.

Grosvenor, Edwin A. "The Races of Europe: The Graphic Epitome of a Never-Ceasing Human Drama. The Aspirations, Failures, Achievements, and Conflicts of the Polyglot People of the Most Densely Populated Continent." *National Geographic Magazine* 34.6 (Dec. 1918): 441–534. Print.

Grünzweig, Walter. "'Collaborators in the Great Cause of Liberty and Fellowship': Whitman as an Intercultural Phenomenon." *Walt Whitman Quarterly Review* 5.4 (1988): 16–26. *Iowa Research Online*. Web. 13 Mar. 2014.

———. "Imperialism." Kummings 151–63. Print.

Gubar, Susan. "The Echoing Spell of H.D.'s *Trilogy*." Friedman and DuPlessis 297–317. Print.

Guest, Barbara. *Herself Defined: The Poet and Her World*. Garden City, NY: Doubleday, 1984. Print.

Gygax, Franziska. *Gender and Genre in Gertrude Stein*. Santa Barbara: Greenwood/Praeger, 1998. Print.

Hadaway, Bridget. "Hammerton, Sir John Alexander (1871–1949)." *Oxford Dictionary of National Biography*. Ed. H. C. G. Matthew and Brian Harrison. Oxford: Oxford University Press, 2004. Ed. Lawrence Goldman. Jan. 2011. Web. 4 Aug. 2012.

Halttunen, Karen. "Transnationalism and American Studies in Place." *The Japanese Journal of American Studies* 18 (2007): 5–19. Print.

Hammerton, Sir John Alexander. "The Editor Explains the Plan and Purpose of This Work." *Countries of the World* 1 (Feb. 1924): i–v. *Conrad First: The Joseph Conrad Periodical Archive*. Web. 4 Aug. 2012.

Harrington, Joseph. "Why Poetry Is Not American Literature." *American Literary History* 8.3 (1996): 496–515. Print.

Harvey, David. "Cosmopolitanism and the Banality of Geographical Evils." *Public Culture* 12.2 (2000): 529–64. Print.

Harvey, Joyce, and Marilyn Ogilvie, eds. *The Biographical Dictionary of Women in Science: Pioneering Lives from Ancient Times to the Mid-Twentieth Century*. Vol. 1. London: Routledge, 2000. Print.

Hawkins, Stephanie L. *American Iconographic:* National Geographic, *Global Culture, and the Visual Imagination*. Charlottesville: University of Virginia Press, 2010. Print.

H.D. *Collected Poems, 1912–1944*. Vol. 2. Ed. Louis L. Martz. New York: New Directions, 1983. Print.

———. "I Sing Democracy." *Life and Letters Today* 17.9 (1937): 154–59. Print.

———. *Paint It Today*. New York: New York University Press, 1992. Print.

———. *Palimpsest*. 1926. Carbondale: Southern Illinois University Press, 1968. Print.

Hegglund, Jon. "Hard Facts and Fluid Spaces: 'Ithaca' and the Imperial Archive." Joyce, Imperialism, and Postcolonialism. Ed. Leonard Orr. Syracuse: Syracuse University Press, 2008. 58-74. Print.

—-. *World Views: Metageographies of Modernist Fiction*. Oxford: Oxford University Press, 2012. Print.

Henry, Barbara. "The Design and Typography of *Leaves of Grass* (1860)." *Huntington Library Quarterly* 73.4 (2010): 601–12. Print.

Hoffman, Tyler. *American Poetry in Performance: From Walt Whitman to Hip Hop*. Ann Arbor: University of Michigan Press, 2011. Print.

Hollenberg, Donna Krolik. *Between History and Poetry: The Letters of H.D. and Norman Holmes Pearson*. Iowa City: University of Iowa Press, 1997. Print.

Holmes, John, ed. *Science in Modern Poetry: New Directions*. Liverpool: Liverpool University Press, 2012. Print.

Honey, Maureen, ed. *Shadowed Dreams: Women's Poetry of the Harlem Renaissance*. New Brunswick: Rutgers University Press, 1989. Print.

Houghton, Walter, ed. *The Wellesley Index to Victorian Periodicals 1824–1900*. 5 vols. Toronto: University of Toronto Press, 1966. Print.

Hoving, Thomas. *Tutankhamun: The Untold Story*. New York: Simon, 1978. Print.

Howes, Marjorie. "'Goodbye Ireland I'm Going to Gort': Geography, Scale, and Narrating the Nation." *Semicolonial Joyce*. Ed. Derek Attridge and Marjorie Howes. Cambridge: Cambridge University Press, 2000. 58–77. Print.

———. "Joyce, Colonialism, and Nationalism." *The Cambridge Companion to James Joyce*. Ed. Derek Attridge. Cambridge: Cambridge University Press, 2004. 254–71. Print.

Hsu, Hsuan L. *Geography and the Production of Space in Nineteenth-Century American Literature*. Cambridge: Cambridge University Press, 2010. Print.

Hubbard, Gardiner Greene. "Africa, Its Past and Future." *National Geographic Magazine* 1.2 (Apr. 1889): 99–124. Print.

———. "Geographic Progress of Civilization: Annual Address by the President Honorable Gardiner G. Hubbard." *National Geographic Magazine* 6 (Feb. 1894): 1–22. Print.

———. "Synopsis of a Course of Lectures on the Effects of Geographic Environment in Developing the Civilization of the World." *National Geographic Magazine* 8.1 (Jan. 1897): 29–32. Print.

Huggins, Nathan. *The Harlem Renaissance*. New York: Oxford University Press, 2007. Print.

Hughes, Langston. *The Big Sea: An Autobiography*. New York: Thunder's Mouth Press, 1986. Print.

———. "Calls Whitman Negroes' First Great Poetic Friend, Lincoln of Letters." *Chicago Defender* 4 July 1953: 11. *ProQuest Historical Newspapers*. Web. 2 Mar. 2014.

———. *The Collected Poems of Langston Hughes*. Ed. Arnold Rampersad and David Roessel. New York: Vintage, 1994. Print.

Hughes, Langston, and Arna Bontemps, eds. *The Poetry of the Negro, 1746–1949*. Garden City: Doubleday, 1949. Print.

Humboldt, Alexander von. *Cosmos: A Sketch of a Physical Description of the Universe*. Vol. 1. Trans. E. C. Otté. London: Henry G. Bohn, 1849. Print.

Huntington, Ellsworth. *The Character of Races: As Influenced by Physical Environment*. New York: Charles Scribner, 1924. Print.

———. *Civilization and Climate*. New Haven: Yale University Press, 1915. Print.

———. *World-Power and Evolution*. New Haven: Yale University Press, 1919. Print.

Hutchinson, George B. *The Harlem Renaissance in Black and White*. Cambridge: Harvard University Press, 1995. Print.

———. "Langston Hughes and the 'Other' Whitman." R. Martin. 16–27.

Jakobson, Roman. "Linguistics and Poetics." *Language in Literature*. Cambridge: Harvard University Press, 1987. 62–95. Print.

Jameson, Fredric. *Postmodernism, or, the Cultural Logic of Late Capitalism*. Durham: Duke University Press, 1991. Print.

Jauss, Hans Robert. *Toward an Aesthetics of Reception*. University of Minnesota Press, 1982. Print.

Johnston, Ron. "Creating Human Geography in the English-Speaking World." *The Wiley-Blackwell Companion to Human Geography*. Ed. John A. Agnew and James S. Duncan. Chichester: Wiley-Blackwell, 2011. 89–113. Print.

Jonas, Andrew. "A New Regional Geography of Localities?" *Area* 20.2 (1988): 101–10. Print.

de Jongh, James. "The Poet Speaks of Places: A Close Reading of Langston Hughes' Literary Use of Place." *Historical Guide to Langston Hughes*. Ed. Steven C. Tracy. New York: Oxford University Press, 2004. 65–84. Print.

Joyce, James. *A Portrait of the Artist as a Young Man*. New York: Penguin, 1964. Print.

Kadlec, David. *Mosaic Modernism: Anarchism, Pragmatism, Culture*. Baltimore: Johns Hopkins University Press, 2000. Print.

Kallis, Aristotle. *Fascist Ideology: Territory and Expansionism in Italy and Germany, 1922–1945*. London: Routledge, 2000. Print.

Keegan, John. *The Second World War*. New York: Penguin, 1989. Print.

Keighren, Innes. *Bringing Geography to Book: Ellen Semple and the Reception of Geographical Knowledge*. London: I. B. Tauris, 2010. Print.

Kellner, [Charlotte] L. *Alexander von Humboldt*. London: Oxford University Press, 1963. Print.

Kingham, Victoria. "'Audacious Modernity': *The Seven Arts* (1916–17); *The Soil* (1916–17); and *The Trend* (1911–15)." *The Oxford Critical and Cultural History of Modernist Magazines*. Vol. II. Ed. Peter Brooker and Andrew Thacker. Oxford: Oxford University Press, 2012. 398–419. Print.

Kish, George. *A Sourcebook in Geography*. Cambridge: Harvard University Press, 1978. Print.

Klammer, Martin. *Whitman, Slavery, and the Emergence of* Leaves of Grass. University Park: Penn State University Press, 1995. Print.

Klinghoffer, Arthur Jay. *The Power of Projections: How Maps Reflect Global Politics and History*. Santa Barbara: Praeger, 2006. Print.

Kloepfer, Deborah Kelly. *The Unspeakable Mother: Forbidden Discourse in Jean Rhys and H.D.* Ithaca: Cornell University Press, 1989. Print.

Koelsch, William A. "Academic Geography, American Style: An Institutional Perspective." Dunbar 245–80.

Kummings, Donald D., ed. *A Companion to Walt Whitman*. Malden, MA, and Oxford: Blackwell, 2006. Print.

Kusch, Celena. "How the West Was One: American Modernism's Song of Itself." *American Literature* 74.3 (2002): 517–38. Print.

Lamothe, Daphne. *Inventing the New Negro: Narrative, Culture, and Ethnography.* Philadelphia: University of Pennsylvania Press, 2011. Print.

Latour, Bruno. *Pandora's Hope: Essays on the Reality of Science Studies.* Cambridge: Harvard University Press, 1999. Print.

Lefebvre, Henri. *The Production of Space.* Trans. Donald Nicholson-Smith. Oxford: Blackwell, 1991. Print.

"Literary Notices." Rev. of *Physical Geography,* by Mary Somerville. *Godey's Lady's Book* Oct. 1853: n. pag. *Accessible Archives.* Web. 5 May 2013.

"Literature: Mental Reconstruction." Rev. of *World Power and Evolution,* by Ellsworth Huntington; *Can Mankind Survive?,* by Morrison I. Swift; *Remaking a Man,* by Courtenay Baylor; *The Child's Unconscious Mind,* by Wilfrid Lay; and *The Mental Hygiene of Childhood,* by William A. White. *Nation* 108.2813 (31 May 1919): 871–72. Print.

Livingston, Robert Eric. "Glocal Knowledges: Agency and Place in Literary Studies." *PMLA* 116.1 (2001): 145–57. Print.

Livingstone, David. *The Geographical Tradition.* Malden, MA, and Oxford: Wiley-Blackwell, 1992. Print.

Loughran, Trish. *The Republic in Print: Print Culture in the Age of U.S. Nation Building, 1770–1870.* New York: Columbia University Press, 2009. Print.

Lutz, Catherine A., and Jane L. Collins. *Reading* National Geographic. Chicago: University of Chicago Press, 1993. Print.

Lynes, Katherine R. "'A Real Honest-to-Cripe Jungle': Contested Authenticities in Helene Johnson's 'Bottled.'" *Modernism/modernity* 14.3 (2007): 517–25. *Project Muse.* Web. 4 Mar. 2014.

———. "'Sprung from American Soil': The 'Nature' of Africa in the Poetry of Helene Johnson." *Interdisciplinary Studies in Literature and Environment* 16.3 (2009): 525–49. *Oxford Journals.* Web. 4 Mar. 2014.

MacGowan, Christopher. "*Paterson* and Local History." *William Carlos Williams Review* 24.2 (2004): 49-60. Print.

Marshall, Alan. "'England and Nowhere.'" *The Cambridge Companion to T.S. Eliot.* Cambridge: Cambridge University Press, 2004. 137-52. Print.

Martin, Geoffrey J. *All Possible Worlds: A History of Geographical Ideas.* New York: Oxford University Press, 2005. Print.

———. "Ontography and Davisian Physiography." Blouet 279–90. Print.

———. Rev. of *Bringing Geography to Book: Ellen Churchill Semple and the Reception of Geographical Knowledge. Geographical Review* 102.1 (2012): 132–35. Print.

Martin, Robert K., ed. *The Continuing Presence of Walt Whitman: The Life after the Life.* Iowa City: University of Iowa Press, 1992. Print.

"Mary Somerville." *Atlantic Monthly* 5 May 1860: 568–71. *Electronic Text Center* at University of Virginia. Web. 16 May 2013.

"Mary Somerville." *Scientific American* 4 Sept. 1869: 150. Print.

Massey, Doreen. *For Space*. London: Sage, 2005. Print.

Matteson, John. "Humboldt, Alexander von (1769–1859)." *The Routledge Encyclopedia of Walt Whitman*. Ed. J. R. LeMaster and Donald D. Kummings. New York: Routledge, 1998. 288–89. Print.

Maxwell, William. "Global Poetics and State-Sponsored Transnationalism: A Reply to Jahan Ramazani." *American Literary History* 18.2 (2006): 360–64. Print.

McGann, Jerome. "How to Read a Book." *New Directions in Textual Studies*. Ed. Dave Oliphant and Robin Bradford. Austin: University of Texas Press, 1990. 13–38. Print.

McGill, Meredith L. *American Literature and the Culture of Reprinting, 1834–1853*. Philadelphia: University of Pennsylvania Press, 2007. Print.

Meyer, Steven. *Irresistible Dictation: Gertrude Stein and the Correlations of Writing and Science*. Stanford: Stanford University Press, 2001. Print.

Mikesell, Martin. "Continuity and Change." Blouet 1–15.

Miller, Meredith. "Enslaved to Both These Others: Gender and Inheritance in H.D.'s 'Secret Name: Excavator's Egypt.'" *Tulsa Studies in Women's Literature* 16.1 (1997): 77–105. Print.

Miller, W. Jason. *Langston Hughes and American Lynching Culture*. Gainesville: University Press of Florida, 2011. Print.

Mitchell, Don. "Public Housing in Single-Industry Towns: Changing Landscapes of Paternalism." *Place/Culture/Representation*. Ed. James Duncan and David Ley. London: Routledge, 1993. 110–27. Print.

Mitchell, Verner D., ed. and "Introduction." *This Waiting for Love: Helene Johnson, Poet of the Harlem Renaissance*. Amherst: University of Massachusetts Press, 2000. 1–20. Print.

Moore, Ralph Westwood. Rev. of *The Geography of the Mediterranean Region: Its Relation to Ancient History*, by Ellen Churchill Semple. *Times Literary Supplement* (London) 21 July 1932: 524. *Times Literary Supplement*. Web. 27 Mar. 2013.

Moretti, Franco. *Signs Taken for Wonders: On the Sociology of Literary Forms*. London and New York: Verso, 2005. Print.

Morris, Adelaide. *How to Live / What to Do: H.D.'s Cultural Poetics*. Urbana and Chicago: University of Illinois Press, 2003. Print.

Mortimer, Raymond. "The Mediterranean." Rev. of *The Geography of the Mediterranean Region: Its Relation to Ancient History*, by Ellen Churchill Semple. *The New Statesman and Nation* 4 (27 Aug. 1932): 237–38. Print.

Moses, Michael Valdez. *The Novel and the Globalization of Culture*. Oxford and New York: Oxford University Press, 1995. Print.

Murray, William. *Janet, My Mother, and Me: A Memoir of Growing Up with Janet Flanner and Natalie Danesi Murray*. New York: Simon and Schuster, 2000. Print.

Natter, Wolfgang, and John Paul Jones, III. "Signposts toward a Poststructuralist Geography." *Postmodern Contentions: Epochs, Politics, Space*. Ed. John Paul Jones III, Wolfgang Natter, and Theodore R. Schatzki. New York: Guilford Press, 1993. 165–203. Print.

Neeley, Kathryn A. *Mary Somerville: Science, Illumination, and the Female Mind.* Cambridge: Cambridge University Press, 2001. Print.

Nelson, Cary. *Repression and Recovery: Modern American Poetry and the Politics of Cultural Memory, 1910–1945.* Madison: University of Wisconsin Press, 1989. Print.

Nicholls, Peter. "Lost Object(s): Ezra Pound and the Idea of Italy." *Ezra Pound and Europe.* Ed. Richard Taylor and Claus Melchior. Amsterdam: Rodopi, 1993. 165–76. Print.

North, Michael. *Reading 1922.* Oxford: Oxford University Press, 1999. Print.

O'Connor, William Douglas. "Walt Whitman." *New York Times* 2 Dec. 1866: 2. *Walt Whitman Online Archive.* Ed. Ed Folsom and Kenneth M. Price. Web. 23 June 2013.

Oppenheim, James. "Poetry—Our First National Art." *Dial* 68.2 (Feb. 1920): 238–42. Print.

Ostriker, Alicia. "No Rule of Procedure: The Open Poetics of H.D." Friedman and DuPlessis 336–51. Print.

Pandora, Katherine. "Popular Science in National and Transnational Perspective: Suggestions from the American Context." *Isis* 100.2 (2009): 346–58. *JSTOR.* Web. 23 May 2011.

Paryz, Marek. *The Postcolonial and Imperial Experience in American Transcendentalism.* New York: Palgrave Macmillan, 2012. Print.

Pauly, Philip. "The World and All That Is in It: The National Geographic Society, 1888–1918." *American Quarterly* 31 (1979): 517–32. Print.

Peet, Richard. "The Social Origins of Environmental Determinism." *Annals of the Association of American Geographers* 75.3 (1985): 309–33. Print.

Perelman, Bob. *The Trouble with Genius: Reading Pound, Joyce, Stein, and Zukofsky.* Berkeley: University of California Press, 1994. Print.

Perloff, Marjorie. *The Poetics of Indeterminacy: Rimbaud to Cage.* Evanston, IL: Northwestern University Press, 1981. Print.

Pollak, Vivian R. *The Erotic Whitman.* Berkeley: University of California Press, 2000. Print.

Pondrom, Cyrena. "An Introduction to the Achievement of Gertrude Stein." *Geography and Plays* by Gertrude Stein. Madison: University of Wisconsin Press, 1993. vii–lv. Print.

Poole, Robert M. *Explorers House:* National Geographic *and the World It Made.* New York: Penguin, 2004. Print.

Popenoe, Paul B. "Costa Rica, Land of the Banana." *National Geographic Magazine* 41.2 (Feb. 1922): 201–20. Print.

Porter, Carolyn. "What We Know that We Do Not Know, Remapping American Literary Studies." *American Literary History* 6.3 (1994): 467–526. Print.

Pound, Ezra. *The ABC of Reading.* New York: New Directions, 1934. Print.

———. *The Cantos of Ezra Pound.* New York: New Directions, 1993. Print.

Pratt, Mary Louise. *Imperial Eyes: Travel Writing and Transculturation*. London: Routledge, 1992. Print.

Price, Kenneth M., and Ed Folsom. *Re-scripting Walt Whitman: An Introduction to His Life and Work*. Malden: Blackwell, 2008. Print.

Quinones, Ricardo J. *Mapping Literary Modernism: Time and Development*. Princeton: Princeton, University Press, 1985. Print.

Radway, Janice. "What's in a Name? Presidential Address to the American Studies Association, 20 November, 1998." *American Quarterly* 51.1 (1999): 1–32. Print.

Rainey, Lawrence. *Institutions of Modernism: Literary Elites and Public Culture*. New Haven: Yale University Press, 1998. Print.

Ramazani, Jahan. "Poetry, Modernity, and Globalization." *The Oxford Handbook of Global Modernisms*. Ed. Mark Wollaeger. Oxford: Oxford University Press, 2012. 288–312. Print.

———. *Poetry of Mourning: The Modern Elegy from Hardy to Heaney*. Chicago: University of Chicago Press, 1994. Print.

———. "A Transnational Poetics." *American Literary History* 18.2 (2006): 332–59. Print.

Rampersad, Arnold. *The Life of Langston Hughes: Volume I: 1902–1941, I, Too, Sing America*. New York: Oxford University Press, 1986. Print.

Reaves, Gerri. *Mapping the Private Geography: Autobiography, Identity, and America*. Jefferson, NC: McFarland, 2001. Print.

Redman, Tim. *Ezra Pound and Italian Fascism*. Cambridge: Cambridge University Press, 1991. Print.

Reeves, Nicholas. *The Complete Tutankhamun: The King, the Tomb, the Royal Treasure*. London: Thames, 1990. Print.

Rev. of *Influences of Geographic Environment*, by Ellen Churchill Semple. *Nation* 93 (21 Dec. 1911): 610. *Book Reviews Digest Retrospective*. Web. 2 Mar. 2013.

Rev. of *Physical Geography*, revised London edition, by Mary Somerville. *North American Review* 73.153 (Oct. 1851): 386–411. *JSTOR*. Web. 17 May 2011.

Reynolds, David S. *Walt Whitman's America: A Cultural Biography*. New York: Vintage, 1995. Print.

Riddel, Joseph. "H.D. and the Poetics of 'Spiritual Realism.'" *Contemporary Literature* 10.4 (1969): 447–73. Print.

———. "H.D.'s Scene of Writing—Poetry as (and) Analysis." *Studies in the Literary Imagination* 12 (1979): 41–59. Print.

Roberts, Leo B. "Traveling in the Highlands of Ethiopia." *National Geographic Magazine* 68.3 (1935): 296–328. Print.

Rodogno, David. *Fascism's European Empire: Italian Occupation During the Second World War*. Cambridge: Cambridge University Press, 2006. Print.

Romer, Isabella. *A Pilgrimage to the Temples and Tombs of Egypt, Nubia, and Palestine in 1845–1846*. Vol. 1. London: Bentley, 1846. Print.

Rothenberg, Tamar Y. *Presenting America's World: Strategies of Innocence in* National Geographic Magazine, *1888–1945*. Aldershot, UK, and Burlington, VT: Ashgate, 2007. Print.

Rowe, John Carlos. *The New American Studies*. Minneapolis: University of Minnesota Press, 2002. Print.

———. *Post-Nationalist American Studies*. Berkeley: University of California Press, 2000. Print.

R.P.L. Review of *The Geography of the Mediterranean Region: Its Relation to Ancient History*, by Ellen Churchill Semple. *America* 47.21 (Aug. 1932): 504. *Humanities International Complete*. Web. 27 Mar. 2013.

Ruprecht, Louis A. "Hellenism on Display." *Journal of Modern Greek Studies*. 15.2 (1997): 247–60. Print.

Sachs, Aaron. *The Humboldt Current: Nineteenth-Century Exploration and the Roots of American Environmentalism*. New York: Penguin, 2007. Print.

Said, Edward. *Orientalism*. London: Routledge, 1978. Print.

Sanderson, Marie. "Mary Somerville: Her Work in *Physical Geography*." *Geographical Review* 64.3 (1974): 410–20. *JSTOR*. Web. 7 May 2011.

Sanna, Ellyn. "Biography of T.S. Eliot." *T.S. Eliot*. Ed. Harold Bloom. Broomall, PA: Chelsea House, 2003. 3-44. Print.

Sauer, Carl. "The Morphology of Landscape." Agnew, Livingstone, and Rogers 296–315.

Scheick, William J. "Death and the Afterlife." Kummings 325–40.

Schulten, Susan. *The Geographical Imagination in America, 1880–1950*. Chicago: University of Chicago Press, 2001. Print.

Schweik, Susan. *A Gulf So Deeply Cut: American Woman Poets and the Second World War*. Madison: University of Wisconsin Press, 1991. Print.

Semple, Ellen Churchill. *American History and Its Geographic Conditions*. Boston: Houghton Mifflin, 1903. Print.

———. *American History and Its Geographic Conditions*. Rev. ed., with Clarence Fielden Jones. Boston: Houghton Mifflin, 1933. Print.

———. "The Anglo-Saxons of the Kentucky Mountains: A Study in Anthropogeography." *Geographical Journal* 17.6 (1901): 561–94. Print.

———. "Emphasis upon Anthropo-Geography in Schools." *Journal of Geography* 3 (1904): 366–74 rpt. *Journal of Geography* 89.5 (1990): 208–10. *Periodical Archives*. Web. 27 Mar. 2013.

———. *The Geography of the Mediterranean Region: Its Relation to Ancient History*. London: Constable, 1931. Print.

———. *Influences of Geographic Environment, on the Basis of Ratzel's System of Anthropo-Geography*. New York: Holt, 1911. Print.

"Semple's *Influences of Geographic Environment: On the Basis of Ratzel's System of Anthropogeography*." Rev of *Influences of Geographic Environment*, by Ellen Churchill Semple. *Nation* 93 (21 Dec. 1911): 610–11. *Book Review Digest Retrospective*. Web. 2 Mar. 2013.

Shattock, Joanne, and Michael Wolff, eds. *The Victorian Periodical Press: Samplings and Soundings*. Leicester: Leicester University Press, 1982. Print.

Sherry, Vincent. "T. S. Eliot, Late Empire, and Decadence." Begam and Moses 111–35.

Showalter, William Joseph. "New York—An Empire within a Republic." *National Geographic Magazine* 64.5 (1933): 513–84. Print.

Silverstein, Louis. "Planting the Seeds: Selections from The H.D. Chronology." *H.D. Newsletter* 2.2 (1988): 4–14. Print.

Simpich, Frederick. "This Giant That Is New York." *National Geographic Magazine* 58.5 (Nov. 1930): 517–83. Print.

Smethurst, James. *The New Red Negro: The Literary Left and African American Poetry, 1930–1946.* Oxford: Oxford University Press, 1999. Print.

Smith, Neil. *American Empire: Roosevelt's Geographer and the Prelude to Globalization.* Berkeley: University of California Press, 2003. Print.

Smitten, Jeffrey R., and Ann Daghistany, ed. *Spatial Form in Narrative.* Ithaca: Cornell University Press, 1981. Print.

Smyers, Virginia. "H.D.'s Books in the Bryher Library." *H.D. Newsletter* 1.2 (1987): 18–25. Print.

Soja, Edward. *Postmodern Geographies: The Reassertion of Space in Critical Social Theory.* London: Verso, 1989. Print.

Sollors, Werner. *Ethnic Modernism.* Cambridge: Harvard University Press, 2008. Print.

Somerville, Mary. *Personal Recollections from Early Life to Old Age.* New York: AMS Press, 1975. Print.

———. *Personal Recollections from Early Life to Old Age.* 3rd ed. London: John Murray, 1851. Print.

———. *Queen of Science: Personal Recollections of Mary Somerville.* Ed. Dorothy McMillan. Edinburgh: Cannongate Classics, 2001. Print.

Stein, Gertrude. "An American in France." *What Are Masterpieces.* Los Angeles: Conference Press, 1940. Print.

———. *Everybody's Autobiography.* Cambridge: Exact Change, 1993. Print.

———. *The Geographical History of America; or, The Relation of Human Nature to the Human Mind.* Baltimore: Johns Hopkins University Press, 1995. Print.

———. *How Writing is Written: Volume II of the Previously Uncollected Writings of Gertrude Stein.* Ed. Robert Bartlett Haas. Los Angeles: Black Sparrow Press, 1974. Print.

———. *Painted Lace and Other Pieces.* New Haven: Yale University Press, 1955. Print.

———. *Useful Knowledge.* London: John Lane Bodley Head, 1929. Print.

———. *Yale Gertrude Stein: Selections.* Ed. Richard Kostelanetz. New Haven: Yale University Press, 1980. Print.

Steinman, Lisa Malinowski. *Made in America: Science, Technology, and American Modernist Poets.* New Haven: Yale University Press, 1987. Print.

Stevens, Harold Ray, and J. H. Stape. "The Essays." *The Cambridge Companion to the Works of Joseph Conrad: Last Essays.* Cambridge: Cambridge University Press, 2010. 202–63. Print.

Stimpson, Catharine R. "Gertrude Stein and the Transposition of Gender." *The Poetics of Gender.* Ed. Nancy K. Miller. New York: Columbia University Press, 1986. 1–18. Print.

Terrell, Carroll F. *A Companion to "The Cantos of Ezra Pound."* Berkeley: University of California Press, 1993. Print.

Terrell, John Edward. "Anthropologizing *Guns, Germs, and Steel.*" *Anthropology News* 45.2 (Feb. 2004): 9–11. Print.

Thomas, M. Wynn. *Transatlantic Connections: Whitman U.S., Whitman U.K.* Iowa City: University of Iowa Press, 2005. Print.

Tichi, Cecelia. *Shifting Gears: Technology, Literature, Culture in Modernist America.* New Haven: Yale University Press, 1987. Print.

Traubel, Horace. *With Walt Whitman in Camden.* 9 vols. Boston: Small, Maynard, 1906. *Walt Whitman Online Archive.* Web. 20 June 2013.

Tuason, Julie A. "The Ideology of Empire in *National Geographic Magazine*'s Coverage of the Philippines, 1898–1908." *Geographical Review* 89.1 (1999): 34–53. Print.

Untermeyer, Louis, ed. *Modern American Poetry: An Introduction.* New York: Harcourt, Brace, and Howe, 1919. Print.

Vanskike, Elliott L. "'Seeing Everything as Flat': Landscape in Gertrude Stein's Useful Knowledge and *The Geographical History of America.*" *Texas Studies in Language and Literature* 35.2 (1995): 151–67. *JSTOR.* Web. 21 Jun 2013.

Vetter, Lara. *Modernist Writings and Religio-Scientific Discourse: H.D., Loy, and Toomer.* New York: Palgrave Macmillan, 2010. Print.

Vidal de la Blache, Paul. *Principles of Human Geography.* Trans. Millicent Todd Bingham. 1926. New York: Holt, 1926. Print.

———. *Tableau de la Geographie de la France.* Paris: Hachette, 1903. Print.

Wadsworth, Sarah. *In the Company of Books: Literature and Its "Classes" in Nineteenth-Century America.* Boston: University of Massachusetts Press, 2006. Print.

Wagner-Martin, Linda. *Favored Strangers: Gertrude Stein and Her Family.* New Brunswick: Rutgers University Press, 1997. Print.

Wald, Priscilla. *Constituting Americans: Cultural Anxiety and Narrative Form.* Durham: Duke University Press, 1995. Print.

———. "Minefields and Meeting Grounds: Transnational Analyses and American Studies." *American Literary History* 9.1 (Spring 1997): 199–218. Print.

Walkowitz, Rebecca. *Cosmopolitan Style: Modernism Beyond the Nation.* New York: Columbia University Press, 2006. Print.

Walkowitz, Rebecca, and Doug Mao. "The New Modernist Studies." *PMLA* 123.3 (2008): 737–48. Print.

Walls, Laura Dassow. *Emerson's Life in Science: The Culture of Truth.* Ithaca: Cornell University Press, 2003. Print.

———. *The Passage to Cosmos: Alexander von Humboldt and the Shaping of America.* Chicago: University of Chicago Press, 2009. Print.

———. *Seeing New Worlds: Henry David Thoreau and Nineteenth-Century Natural Science.* Madison: University of Wisconsin Press, 1995. Print.

Walsh, Rebecca. "Where Materiality Meets Metaphor: The Spatialized Subject and the Limits of Locational Feminism." *Exclusions in Feminist Thought: Challenging the Boundaries of Womanhood.* Ed. Mary F. Brewer. Brighton: Sussex Academic Press, 2002. 182–204. Print.

Wand, David Hsin-Fu. "To the Summit of Tai Shan: Ezra Pound's Use of Chinese Mythology." *Paideuma* 3 (1974): 3–12. Print.

Warmington, Eric H. *Greek Geography*. London: J. M. Dent, 1934. Print.

Warren, James P. "Reading Whitman's Postwar Poetry." *The Cambridge Companion to Walt Whitman.* Ed. Ezra Greenspan. Cambridge: Cambridge University Press, 1995. 45–65. Print.

Watten, Barrett. "Method and Surrealism: The Politics of Poetry." *LANGUAGE* 4 (1982): 129–40. Print.

Wertheimer, Eric. *Imagined Empires: Incas, Aztecs, and the New World of American Literature, 1771–1876*. Cambridge: Cambridge University Press, 1999. Print.

Westover, Jeff. "Africa/America: Fragmentation and Diaspora in the Work of Langston Hughes." *Callaloo* 25.4 (2002): 1206–23. *JSTOR*. Web. 13 Jun 2013.

"What to Read." *Crisis* 2.4 (Aug. 1911): 168. Print.

"What to Read." *Crisis* 3.1 (Nov. 1911): 34. Print.

Whitbeck, R. H. "Obituary: Ellen Churchill Semple." *Geographical Review* 22.3 (1932): 500–501. Print.

Whitman, Walt. *Leaves of Grass and Other Writings*. Ed. Michael Moon. New York: Norton, 1992. Print.

———. "*North British Review*, May 1844." Lincoln Material Poetry Manuscripts "The Crusades" [1869?]. *Walt Whitman Online Archive*. Ed. Ed Folsom and Kenneth M. Price. Web. 25 May 2011.

Whitworth, Michael H. *Einstein's Wake: Relativity, Metaphor, and Modernist Literature.* Oxford: Oxford University Press, 2001. Print.

———. "'Strange Synthetic Perfumes': Investigating Scientific Diction in Twentieth-Century Poetry." Holmes 87–100. Print.

Williams, Raymond. "When Was Modernism?" *The Politics of Modernism*. Ed. Tony Pinkney. London: Verso, 1989. 31–36. Print.

Williams, William Carlos. *Paterson*. 5 vols. New York: New Directions, 1992. Print.

Willinsky, John. *Learning to Divide the World: Education at Empire's End.* Minneapolis: University of Minnesota Press, 1998. Print.

Willis, Patricia C. *Marianne Moore: Vision into Verse.* Philadelphia: Rosenbach Museum and Library, 1987. Print.

Witte, Sarah. "The Archaeological Context of H.D.'s 'Secret Name' and 'Hesperia.'" *Sagerieb* 15.1–2 (1996): 51–68. Print.

Woods, Gregory. "'Still on My Lips': Walt Whitman in Britain." R. Martin 129–40.

Wrigley, E. Anthony. "Changes in the Philosophy of Geography." *Frontiers in Geographical Teaching*. Ed. R. J. Chorley and P. Haggett. London: Methuen, 1965. 3–24. Print.

Zilboorg, Caroline. *Richard Aldington and H.D: Their Lives in Letters, 1918–61.* Manchester: Manchester University Press, 2003. Print.

Index

Page numbers in *italics* refer to illustrations.

Rebecca Walsh is assistant professor of English at North Carolina State University, where her research and teaching focus on transnational modernism, American Studies, and world literature. She has published essays in edited volumes, coauthored an essay in *PMLA,* and guest-edited the special issue, "Global Diasporas," of *Interventions: International Journal of Postcolonial Studies.*